DIFFRACTION GRATINGS

Techniques of Physics

Editors

N. H. MARCH

Department of Theoretical Chemistry, University of Oxford, Oxford, England

H.N. DAGLISH

British Telecom Headquarters, 151 Gower Street, London, England

Techniques of physics find wide application in biology, medicine, engineering and technology generally. This series is devoted to techniques which have found and are finding application. The aim is to clarify the principles of each technique, to emphasize and illustrate the applications and to draw attention to new fields of possible employment.

1. D. C. Champeney: Fourier Transforms and their Physical Applications.

2. J. B. Pendry: Low Energy Electron Diffraction.

3. K. G. Beauchamp: Walsh Functions and their Applications.

4. V. Cappellini, A. G. Constantinides and P. Emiliani: Digital Filters and their Applications.

5. G. Rickayzen: Green's Functions and Condensed Matter.

6. M. C. Hutley: Diffraction Gratings.

DIFFRACTION GRATINGS

M.C. HUTLEY

National Physical Laboratory
Teddington, Middlesex

1982

ACADEMIC PRESS

A Subsidiary of Harcourt Brace Jovanovich, Publishers

London   New York
Paris   San Diego   San Francisco
São Paulo   Sydney   Tokyo   Toronto

Academic Press Inc. (London) Ltd.
24–28 Oval Road
London NW1

US edition published by
Academic Press Inc.
111 Fifth Avenue,
New York, New York 10003

British Library Cataloguing in Publication Data
Hutley, M. C.
  Diffraction gratings. — (Techniques of physics,
  ISSN 0308-5392; 6)
  1. Diffraction gratings
  I. Title II. Series
  681'.4   QC417   LCCCN 81-6783

  ISBN 0-12-362980-2

Printed in Great Britain at the Alden Press, Oxford

Preface

Diffraction gratings have been known for over 150 years. Because of their importance as the dispersing elements in spectroscopic instruments, they have been the subject of considerable and sustained research. The design and construction of an engine for ruling gratings is one of the most demanding tasks that can be asked of an engineer. Thanks to the dedicated efforts of ruling engineers since the end of the last century, there have been constant improvements in quality and as high-quality replicas became more widely available, so gratings came more and more to replace prisms. By the mid 1960s there was a considerable body of knowledge and expertise concerning the manufacture and testing of gratings and the designing of instruments using ruled gratings. However, since then there have been two dramatic developments which have added a new dimension to the subject. The first is the development of interference (or so called "holographic") techniques of manufacture and the second is the use of powerful computers to solve Maxwell's equations and thereby to describe fully the behaviour of gratings.

In writing this book I have sought to do three things. The first is to set down the basic facts about gratings in slightly greater depth than is to be found in most text books on optics and to include the practical aspects of the manufacture, testing and use of gratings in addition to the basic theory. The second is to summarize some recent developments and to put them into perspective with respect to the established state of the art. Finally I have taken the opportunity to describe some of those aspects of the subject that I have found particularly fascinating. The book is therefore something of a mixture of a text book, a review article and a personal anthology and as a consequence the reader may find some inconsistencies in the level of prior knowledge assumed. However, as far as possible I have avoided detailed mathematics in the hope that the text will be acceptable to anyone with a scientific background. I hope in particular that it may help the spectroscopist to appreciate the way in which the properties and practical limitations of the grating influence the performance of his instrument, and I hope that it may help him to use to the best advantage the gratings which are available.

The subject of gratings is both broad and deep and it is unavoidable that some aspects have to be treated somewhat superficially. It may therefore be pertinent to point out some of the things that this book does *not* seek to do. It does not attempt a detailed description of the development of the ruling engine, but rather a summary of the principles that are involved. It does not attempt a detailed description of the electromagnetic theory, but again a summary is given of the principles and of the main results. (Further details of these topics may be found in the works of Stroke (1967) and of Petit (1980) respectively). Finally it should be noted that the subject of the book has been limited to that of gratings, mostly but not exclusively spectroscopic gratings, and it therefore does not deal with the design of spectroscopic instruments as such, or with the subject of spectroscopy even though many of the recent advances in these subjects are largely due to the development of new and better gratings.

I would like to record my thanks to all who have contributed to the preparation of this book, either directly by reading and correcting parts of the text, or indirectly by passing on to me the benefit of their knowledge and experience. In particular I would like to thank Mr J. F. Verrill, Mrs J. Wilson, Mrs J. Kirby and Mr R. F. Stevens for their valuable help.

M. C. HUTLEY
National Physical Laboratory
October 1981

Contents

Introduction

██

Whenever a travelling wave encounters an obstruction with dimensions similar to its wavelength, some of the energy in the wave is scattered. If the obstruction is periodic, or indeed if there is a periodic variation of any parameter which affects the propagation of the wave, energy is scattered into various discrete directions or "diffracted orders", and a structure which acts in this way may be referred to as a "diffraction grating". From the point of view of a wave in a diffracted order, the effect of the grating is to change the direction of propagation, and the amount by which it does so depends upon the relationship between the wavelength and the period. In this way a grating may disperse a variety of wavelengths to form a spectrum. It therefore performs the same function as a prism, but in many respects it does so better and more conveniently.

In principle any periodic structure may, by interacting with a suitable wave motion, generate diffracted orders. Similarly any wave motion may be split up into a spectrum provided that it interacts with a suitable grating. Thus radio waves can interact with waves on the surface of the ocean and there are occasions when this must be taken into account in navigational instruments. Crystals are periodic arrays of atoms and their diffraction of X-rays and electrons is the basis for practically the whole of the science of crystallography. Diffraction effects can be seen in nature and in the home: they are responsible for the sheen of some bird feathers, for the colours that are observed on gramophone records, and for the coloured patterns that are seen when a street light is viewed through the fabric of an umbrella. The tale is even told of a professor who convinced the magistrate that diffraction by a corrugated iron fence was responsible for an erroneous reading of the radar speed trap in which he had the misfortune to become ensnared. However, the real importance of gratings lies in their ability to produce a spectrum, particularly of electromagnetic radiation, and the fact that by the detailed analysis of this spectrum one can study either the source of the radiation or the medium through which it passed. The term "spectrum" refers either to the distribution of the radiation among the various wavelengths or to the display of that distribution. The basis of spectroscopy is the separation and ordering of the various wavelengths that are present, in order to display them as a spectrum. In cases such as the emission of light from atoms in a discharge there are comparatively few discrete wavelengths present, whereas in others — such as the emission from a "black body" — all wavelengths are present and we have a continuous spectrum or "white light".

There are two categories of spectroscopic instruments, the spectrometer which serves to measure or display a spectrum and the monochromator which selects and transmits a single wavelength or a narrow band of wavelengths from the spectrum. Instruments which employ gratings separate the different components of the spectrum by redirecting the radiation by an amount which depends upon wavelength. In order to avoid confusion the radiation must approach the dispersing element (which could be a grating or a prism) from a well defined direction and this is achieved by passing it through a slit and a collimator. After the radiation has been dispersed it is brought to a focus so that the spectrum is displayed as a series of images of the slit each positioned according to its wavelength. In the case of an emission spectrum where there are discrete wavelengths present these are seen as a series of images of the slit or "spectral lines". In a continuum there is an infinite number of lines which merge into each other.

Each atom and each molecule has its own characteristic spectrum by which it may be identified and spectroscopy (which is the analysis of the spectrum) is a science which has a very wide range of applications. It is used, for example, in hospitals in the analysis of blood samples, and it is used industrially in both process control and quality control. It checks the constituents of the melt in the production of steel and it detects minute quantities of impurities in all manner of products. In research the applications of spectroscopy are amazingly diverse. On the one hand it enables us to study the innermost features of the atom, while on the other it is our only means of finding out what the stars are made of. At the heart of most of this work lies the diffraction grating. In many spectroscopic instruments the grating accounts for only a small proportion of the cost, but is the quality of the grating which sets the ultimate limit on the quality of the spectrum that can be obtained. It is this that makes the diffraction grating so important and it is this type of grating that will be the subject of this book. We shall consider how various features of a grating influence the performance of the instrument in which it is used, how over the past century there has been a continual struggle to improve the quality of gratings, and why it is that the struggle still continues.

Although any periodic structure will act as a diffraction grating, we shall restrict ourselves mainly to the consideration of a one-dimensional array that is in effect a series of parallel equispaced lines of one form or another lying in the same plane. Any form of wave motion can be diffracted by a suitable grating, but it is in the field of optics that gratings are most important and have been developed to their highest degree of perfection. Much of what follows will apply equally to the diffraction of sound waves or water waves, but it is the gratings for electromagnetic radiation that are our main concern, particularly those for the spectrum between the infrared and X-rays. In practice most gratings consist of a suitable rigid substrate with an optical surface upon which are produced a series of equispaced parallel grooves. Such a grating is shown schematically in Figure 1.1.

The invention of the diffraction grating is ascribed to the American astronomer David Rittenhouse who, in 1786 had been intrigued by the effects produced when viewing a distant light source through a fine silk handkerchief. In order to repeat the phenomenon under controlled conditions, he made up a square of parallel hairs laid across two fine screws made by a watchmaker. When he looked through this at a small opening in the window shutter of a darkened room, he saw three images of approximately equal brightness and several others on either side "fainter and growing more faint, coloured and indistinct, the further they were from the main line". He noted that red light was bent more than blue light and ascribed these effects to diffraction. He never got as far as measuring the wavelength of light, but he did correctly predict that by pursuing these studies, "new and interesting discoveries may be made concerning this interesting substance, light".

The phenomenon of spectral colours produced by a grating was exploited somewhat frivolously in 1822 by Sir John Barton who was

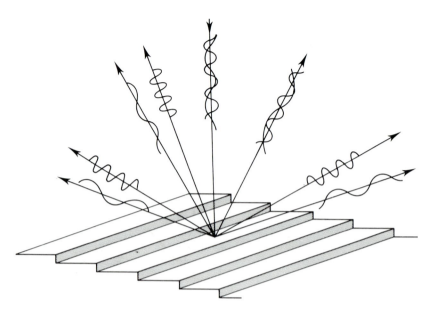

Figure 1.1 The separation of a wave motion into different spectral orders by a diffraction grating.

deputy controller of the Mint in London. He took out a patent entitled "A certain process for the application of prismatic colours to the surface of steels and other metals and using the same in manufacture of various ornaments". In this he claims to exploit the spectral decomposition of light by cutting crossed gratings on steel using a diamond. The purpose of this exercise was to make a die for moulding fancy metal waistcoat buttons for the well dressed man about town of his day.

However, it is Joseph von Fraunhofer who deserves the credit for the invention of the diffraction grating as we know it today. Quite independently in 1821 he repeated Rittenhouse's experiments with fine wire gratings and produced reflection gratings by ruling grooves with a diamond point on a mirror surface. His finest grating was only 12 mm wide and contained 9600 grooves, but this enabled him to conduct a thorough study of its performance and to measure, for the first time, the wavelength of light. He went on to explain the phenomenon of diffracted orders, derived and verified the grating equation and considered the effects on the spectrum of the form of the grooves and of errors in their position. He laid the foundations of a study which was to continue for over 150 years and is still in progress.

Fraunhofer died in 1826 and for the next half a century little interest was shown in gratings. The problem was how to make a grating which would not only produce a coloured spectrum but which would have a performance equal to that of a prism. It soon became apparent that a grating consisting of grooves ruled on a mirror surface was more practical than one consisting of parallel wires and it was at this point that both the English and German nomenclature started to go a little awry. A series of parallel wires could genuinely be said to resemble a miniature grating (or a trellis or picket fence; *Gitter* in German) but this is rather less true of a series of grooves formed in the surface of a mirror.

Initially the manufacture of gratings was undertaken by watch-makers because they were the only people who had the experience of working to the fine scale and high precision demanded. Indeed it is believed that the dividing engine for ruling Barton's buttons was made by W. Harrison, son of J. Harrison who invented the marine chronometer. However, towards the end of the nineteenth century Lord Rayleigh showed, theoretically, that the ability of a grating to resolve spectral lines exceeded that of a prism (Rayleigh 1874). In 1882 H. A. Rowland at the Johns Hopkins University in Baltimore built what he called a "ruling engine" and produced gratings of a quality which proved this to be true. Rowland is gener-ally regarded as the originator (if not technically the inventor) of the ruling engine. This is a device for generating two orthogonal motions, the comparatively rapid motion of a diamond point as it burnishes a groove and a much slower motion perpendicular to the grooves as the diamond is moved from one groove to the next. The tolerances required for an optical grating are generally of the order of a small fraction of the wavelength of visible light and the ruling engine may be regarded as the most precise form of machine tool in existence. Not only did Rowland build the first really success-ful ruling engine but he also invented the concave grating (Rowland 1883) in which the grooves are ruled on the surface of a concave spherical mirror. Such a grating not only disperses light into a spec-trum, but also focuses it into a sharp image. This made possible a wide range of new instrumental designs and greatly extended the range of wavelengths which could be studied.

The work at Baltimore was pursued with great success by J. A. Anderson, by R. W. Wood and by J. Strong, and up until the Second World War the Johns Hopkins University was the world's principal supplier of diffraction gratings. R. W. Wood (1910) developed a tech-nique for controlling the distribution of light among the diffracted

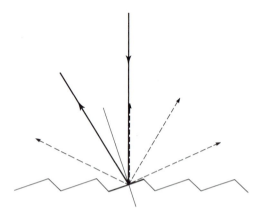

Figure 1.2 The concentration of energy into a chosen order by a blazed grating.

orders by the shape of the grating grooves. In this technique, known as "blazing", the grooves have a sawtooth profile and the facets of the grooves may be considered as a series of small mirrors which tend to reflect the light into the direction of a chosen diffraction order, thereby enhancing the proportion of energy in that order (Figure 1.2). Strong (1935) introduced the vacuum deposition of thin metal films and this enabled gratings to be ruled in a thin layer of a metal such as aluminium on the surface of a glass blank, rather than directly onto a blank of speculum metal. Not only were the new blanks easier to work and easier to obtain, but the reflectance of aluminium is higher than that of speculum and so the gratings were more efficient.

Meanwhile, in England, Otto Hilger built an engine for Lord Blythswood which was commissioned in 1908 and was later transferred to the National Physical Laboratory. There it ruled gratings until it was donated to the Science Museum in 1958. A. A. Michelson at the University of Chicago made notable contributions to the art of ruling gratings, including the suggestion of controlling the position of the grooves by means of an interferometer (Michelson 1915). However, he would be satisfied with nothing short of perfection and so in fact produced very few gratings. The one grating which did appear to satisfy him he subsequently dropped and broke at a dinner party. This unhappy event appears to have earned him as much fame in grating circles as have his considerable technical achievements, but his work was by no means wasted. His first engine was later taken over by the Bausch and Lomb Optical Company of

Rochester, New York. Largely as a result of this they developed an interest in diffraction gratings, turned Michelson's technical successes to commercial advantage and went on to become the world's major supplier. This engine, although modified and updated, is still used for the commercial production of gratings. Michelson's second engine was developed further at the Massachusett's Institute of Technology and became known as the MIT "A" engine.

In all the early ruling engines, and in many of the later ones, the position of the grooves was determined by means of a lead screw and the quality of the grating depended on the accuracy of the screw and of the bearings. Defects in the screw such as ellipticity, eccentricity or drunkenness gave rise to periodic errors in the position of the grooves and this in turn generated spurious images known as ghosts. A great deal of effort went into the reduction of ghosts and into the construction of special cams which would compensate for the errors of the lead screw. However, the fundamental limit to the precision of a ruling engine was the stiffness and stability of the material of which it was made. When one is concerned with displacements of the order of a few atomic diameters, an engine made of the stiffest material behaves as though it were made of rubber. This barrier to progress was overcome by Harrison and Stroke (1955) when they put into practice Michelson's suggestion of a servo system controlled by an interferometer. The positions of the grooves were then determined relative to the wavelength of light; the screw which had been so critical in earlier engines served only to translate the blank perpendicular to the direction of the grooves.

A rather different approach to the problem of lead screw errors was suggested in 1950 by Sir Thomas Merton and was put into practice at the National Physical Laboratory by a team led by L. A. Sayce (1953). In this the grating was produced on a lathe in the form of an accurate helix in a two-stage process. The first stage involved cutting a helix on a brass cylinder using an accurate lathe. This helix preserved the errors of the lead screw of the lathe but was then used on a second machine to generate a further helix which was free from these errors. The second machine was a chasing lathe with a resilient pith nut that spanned many of the grooves of the first helix and averaged out their errors. The technique had the advantage that it avoided the intermittent motion of both the diamond and the blank in the "flat bed" ruling engine, but it had the enormous disadvantage that the grating was on a cylindrical rather than a plane surface. In order to transfer the grating to a plane a plastic pellicle was cast on the helix. This was then split open and a replica of it was cast in gelatine on an optical flat. When the gelatin had hardened, the pellicle was removed and the replica coated with aluminium. The optical quality of such gratings was limited by the distortion of the plastic pellicle, but nevertheless it was possible to produce gratings of good quality for use with infrared radiation. Their availability and low cost contributed significantly to the development of infrared spectroscopy in the 1950s.

Another fundamental problem in the ruling of gratings is the wear and fracture of the diamond ruling tool. The total length of groove in a grating may, on occasions, exceed 30 km and in order to maintain uniformity of efficiency it is necessary that there should be no detectable change in shape of the diamond. This problem is particularly important for the large gratings required for high resolution spectroscopy. However, in 1949 Harrison pointed out that the resolution of a grating does not depend upon the number of grooves but upon the width of the grating. His solution to the problem of diamond wear was to rule comparatively coarse gratings (for example 79 grooves mm^{-1}) and to use them in high diffracted orders at large angles of incidence. Over the next quarter of a century his group at M.I.T. built a series of three ruling engines and produced gratings of the highest resolution possible. The use of this type of grating, known as an "echelle" entails the penalty that a coarse grating gives rise to a large number of diffracted orders and that the spectra in the different orders overlap. To disentangle the spectrum it is necessary to introduce dispersion in an orthogonal direction with a second grating or a prism, but this is a price that many spectroscopists have been prepared to pay.

One of the most important developments in grating technology, second perhaps only to the invention of the ruling engine itself, was the development by White and Fraser in the late 1940s of techniques for high-quality replication. In this technique a non-adherent coating is vacuum-deposited onto the surface of the grating and is followed by a reflective coating, usually of aluminium. Onto this is poured a pool of resin and onto the resin is lowered a glass blank with a surface that has usually been treated to improve its adhesion to the resin. When the resin has cured, the sandwich is separated by force to yield a replica of the grating in resin on the glass blank. The technique is now widely used to the extent that it is taken for granted by spectroscopists. Nevertheless, it remains a rather brutal way of treating a particularly delicate piece of optics that has been produced at enormous expense. The significance of replication is not that it is a means of improving the performance of a grating (although under some circumstances it can do this), but that it makes gratings far more widely available. Despite the great advances in both the understanding and the production of gratings that had been achieved since the beginning of the century, it was not until the 1950s that gratings began to displace prisms in commercial spectroscopic instruments. The delay was not due to inadequacy of the gratings, but because prior to replication they were not available in large numbers. Their availability as replicas led to improvements in spectroscopy and this led in turn to a demand by spectroscopists for gratings of still higher quality. At one time the spectroscopist was lucky if he was able to obtain a precious grating at all, and if it had some fault then he had to make the best of it. Nowadays, the user expects perfection and will often reject a grating with some minor flaws, even if for his purposes it will not affect the grating's performance.

As with many branches of optics, spectroscopy was greatly affected by the invention of the laser in the early 1960s. The properties of laser light, that it is highly monochromatic and that it propagates in a well defined beam, greatly simplified Raman spectroscopy. This led to a renewed interest and a far wider use of it as an analytical technique. The essential feature of Raman spectroscopy is that a small proportion of the light scattered from a sample undergoes a shift in wavelength caused by the internal motions of the sample. The spectra takes the form of a series of very weak spectral lines in the presence of a very much stronger line. Their study requires gratings with extremely low levels of stray light and an absence of ghosts, and so stringent is the requirement for spectral purity that it is customary to use instruments with two or even three gratings in series.

If the laser generated new demands on the quality of gratings required for spectroscopy, it also offered, at least in part, a solution to the problems of their manufacture. It made possible their production by an entirely different technique using the phenomenon of optical interference. When two coherent beams of light intersect they generate a series of interference fringes. These may then be recorded in a photosensitive material and used to form the grooves of the grating. The spacing of the fringes is determined by the angle of intersection of the beams and by the wavelength of light, and since the position of the grooves is determined by the conditions of interference, a grating made in this way will be free from the random and periodic errors present in a grating from a ruling engine.

The origin of the idea of making gratings in this way is obscure, but it was certainly considered by Michelson in 1915. Gratings were made in this way by Burch (1960) but it was not until high-powered lasers were available that it was possible to make gratings which were suitable for general spectroscopic use. This was achieved more or less simultaneously by Rudolph and Schmahl (1967) in Germany and by Labeyrie and Flamand (1967, 1969) in France. Their work coincided with the practical realization of holography, to which it bears some resemblance in that both involve recording interference fringes and that both were made practicable by the invention of the laser. At that time holography was very much in vogue and gratings made in this way were termed "holographic" gratings. For the moment, however, we shall refer to them as interference gratings and reserve the term holographic grating for a more restricted use.

In the production of interference gratings a suitable glass blank is first coated with a thin layer of photoresist which is then exposed to the interference fringe pattern generated by a laser in a suitable interferometer. The relevant property of a photoresist is that its solubility (in a suitable developer) changes with exposure to light, so that the sinusoidal variation of intensity in the fringe pattern is recorded as a sinusoidal variation of solubility, and that upon development this gives rise to a sinusoidal corrugation in the surface of the resist. This process yields a transparent grating in the photoresist which may, if required, be used in transmission, but it is usually coated with a reflecting metal such as aluminium or gold and used in reflection. The technique is simpler, faster and easier to set up than the ruling of gratings, but what is more important from the point of view of the spectroscopist is that the gratings produced are different in many respects. The level of stray light can be very much lower, ghosts are absent, and the spectrum is usually purer than that of even the best ruled gratings. One is not restricted to plane or spherical substrates; nor is one restricted to straight equi-spaced grooves, as other forms of fringe pattern may be generated by altering the shape of the wavefronts in the interferometer. This permits us, for example, to correct holographically for the aberrations of the optical system of the spectrometer and even to design entirely new instruments incorporating gratings of novel shape and with new focusing properties.

However, the interference technique has various shortcomings, the most serious of which is that it affords comparatively little control over the groove profile. In its simplest form it produces gratings with a sinusoidal or a quasi-sinusoidal groove profile.

Although under some circumstances this can be very efficient, it is not always so and there are some gratings which (at the time of writing) the system simply cannot produce. Attempts to make interference gratings with a blazed groove profile have met with some success, particularly for the ultraviolet and X-ray region of the spectrum, but for many applications the most appropriate grating to use is still a ruled one. Interference gratings have not, therefore, replaced ruled gratings but have complemented them and greatly extended the range that is available. There are regions of overlap, Raman spectroscopy and UV–visible spectrophotometry for example, where ruled gratings are being displaced by interference gratings and work is in progress to try to extend this region.

The spectroscopist is now presented with a bewildering choice of gratings. Part of the purpose of the following chapters is to lead him through this choice and to study the advantages and shortcomings of different types of gratings and how these affect the performance of his instrument, with the ultimate aim that he might use it to the best advantage.

2

Theory

☐ The propagation and interaction of waves

In the introduction it was stated that the function of a diffraction grating is to interact with a wave in such a way that it generates a series of further waves, travelling in different directions which are dependent upon the wavelength. Before studying in more detail just how this is achieved, it will be necessary to consider some general features of the propagation and interaction of waves. The features of an ideal plane wave are shown diagrammatically in Figure 2.1.

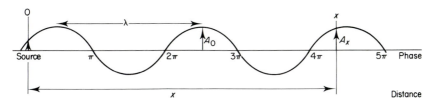

Figure 2.1 The parameters specifying a wave.

At any given point at a distance x from the origin at any time t the amplitude A_x can be represented by the expression

$$A_x = A_0 \sin 2\pi[\nu t - (x/\lambda)] \tag{2.1}$$

where ν is the frequency of oscillation of the source, λ is the wavelength and A_0 is the maximum amplitude of the wave at the source.

13

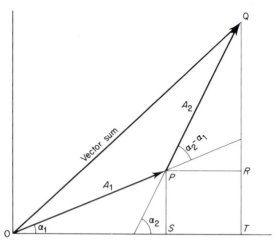

Figure 2.2 The addition of amplitudes.

We may represent the disturbance due to the wave at any point x as a vector of amplitude A_0 and phase $2\pi[\nu t - (x/\lambda)]$. The function $2\pi\nu t$ is the phase of the wave at the source assuming that the amplitude was zero at some time $t = 0$. The second function $- 2\pi(x/\lambda)$ represents the phase lag due to the fact that the wave has travelled from the source to x. The disturbance at a point due to the effect of two or more waves may be determined by the vector addition of the disturbances due to each of these waves, as illustrated in Figure 2.2. Vector 1 has an amplitude A_1 and a phase α_1 and is represented by the line OP, vector 2 has an amplitude A_2 and a phase α_2 and is represented by PQ, and the resultant vector is represented by OQ. The law of cosines gives the length of $(OQ)^2$ as

$$OP^2 + PQ^2 - 2OP \cdot PQ \cos \widehat{OPQ}$$

which is equal to

$$A_1^2 + A_2^2 + 2A_1 A_2 \cos (\alpha_1 - \alpha_2)$$

since $\cos \widehat{OPQ} = - \cos (\alpha_1 - \alpha_2)$. The phase angle of the new vector is $\widehat{QOT}$ of which the tangent is equal to

$$\frac{QT}{OT} = \frac{A_2 \sin \alpha_2 + A_1 \sin \alpha_1}{A_1 \cos \alpha_1 + A_2 \cos \alpha_2}$$

In optics we are often confronted with the problem of determining the irradiance at a point which is illuminated simultaneously by a great many waves. The irradiance is the amount of energy carried by the wave through a unit area in a unit of time and is represented simply by the square of the amplitude of the disturbance. To calculate the resultant irradiance we must, therefore, first determine the resultant amplitude and then square it.

The question of the propagation of waves was considered by Huygens who suggested that each point on a "wave front" should be considered as a new source of waves. A "wave front" is a locus of points at which the disturbance has the same phase and this idea is shown graphically in the left-hand side of Figure 2.3. From this we can see how it is that a new wavefront is built up by contributions from spherical waves generated at each point on the previous wavefront.

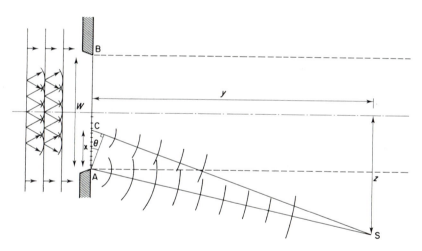

Figure 2.3 Diffraction by an aperture.

If we accept that each point on any wavefront may be considered as a new source radiating in all directions, then we can calculate the amount of light due to this wave that would be detected at any point in space. Consider a point which is not directly illuminated by the beam. It lies in the shadow of an aperture, as shown in Figure 2.3. Let the width W of the beam be defined by a slit-shaped aperture and let the point S be situated at a distance y from the slit and z from the centre of the beam. We consider the wavefront at the aperture to consist of a large number of small sources; each is of equal brightness, each is in phase and each generates a wave of amplitude a which is detected at point S. The total disturbance at the point S is calculated from the vector addition of the disturbances at S due to all these waves. For simplicity let us assume that the slit is infinite in length (that is, into and out of the plane of the paper), so that the diagram represents a typical cross-section of the slit and the waves, and that the secondary wave fronts are in fact cylindrical. The amplitude of the contribution at S for each secondary source at the aperture AB will be equal, but its phase will depend upon the distance that it has had to travel. If we take point A as our reference then the contribution from C will have a phase difference equal to $2\pi(\text{CS} - \text{AS})/\lambda$. As we progress from point A to point B, so the phase difference increases and the vector diagram for the resultant amplitude has the form shown in Figure 2.4. If S is sufficiently far away from the slit that CS and AS are effectively parallel, then the phase angle δ at point C is given by

$$\delta = \frac{x \sin 2\pi\theta}{\lambda} \tag{2.2}$$

when $x = \lambda/2 \sin \theta$ the phase angle is equal to π and the contribution from this point is exactly out of phase with that from A. At this stage the vector diagram approximates to a semicircle.[†]

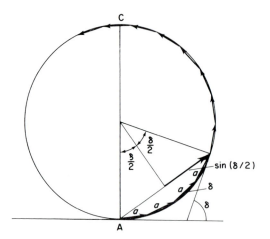

Figure 2.4.

[†] In fact, due to the so-called obliquity factor the contributions decrease in amplitude as $(1 + \cos \theta)$ and the resultant vector traces out a spiral.

At any point the *total* amplitude is proportional to the length of arc of the circle while the *resultant* amplitude is proportional to the chord of the circle. We see from the diagram that the resultant amplitude at S for the whole beam is given by

$$A_S = A_{total} \cdot \frac{2 \sin (\delta/2)}{\delta} = A_{total} \frac{\sin \beta}{\beta} \qquad (2.3)$$

where $A_{total} = \Sigma a$; $(A_{total})^2 = I_0$, the irradiance of the incident wave; β is the phase difference between contributions from the edge and the centre of the beam. It follows that at a point P which is in the shadow the maximum resultant disturbance is represented by the diameter of the circle and is that equivalent to a width $\lambda/2 \sin \theta$ of the aperture. If the width of the beam is much greater than the wavelength, then β is large and the resultant intensity is very small indeed. This analysis shows that even in the region of the geometrical shadow there will be a finite amount of energy. It is due to the phenomenon of *diffraction* and is a consequence of the wave nature of light.

However, it is most important to realize that the reason the irradiance in the shadow is often very small is not that the waves fail to propagate there but that when they do they interfere in such a way that the contributions from most parts of the aperture cancel each other out. If the phase or the amplitide (or both) of the contributions from different parts of the aperture are modified it is possible to reduce the extent to which destructive interference takes place and the diffraction pattern may be modified in such a way that there is a substantial resultant amplitude in what would otherwise have been the shadow.

It is the function of a diffraction grating to provide the appropriate modification to the incident wavefront to generate new wavefronts in a series of well defined directions. Figure 2.5 illustrates three of the ways in which this may be achieved. Consider first what happens if we simply cut out each return half-cycle of the vector diagram. Starting from point A we reach the top of the diagram at a point B corresponding to a point $\lambda/2 \sin \theta$ across the aperture and the resultant contribution from the section AB has an amplitude A_{total}^{AB}/π. The next section of aperture of length $\lambda/2 \sin \theta$ also gives rise to an amplitude of A_{total}^{AB}/π, but out of phase with the first so that it will cancel. This, however, can be eliminated by making that section of the aperture opaque. The next section of the same length will be the same as the first and its amplitude will add, and if we continue across the aperture with alternate transparent and opaque strips then we can see from Figure 2.5(a) that the resultant amplitude will be significant. If the aperture is $N \times (\lambda/2 \sin \theta)$ wide then the intensity will equal

$$\left[\frac{N}{2} \cdot \frac{A_{\text{total}}^{\text{AB}}}{\pi} \right]^2 = I_0 \frac{1}{4\pi^2} \tag{2.4}$$

and we will have generated a so-called "diffracted order" which has resulted from the periodic nature of the aperture. The period d of the grating is equal to the width of two strips, one transparent and one opaque, and thus we can write

$$d = \lambda/\sin \theta \quad \text{or} \quad \lambda = d \sin \theta \tag{2.5}$$

In this example the grating consists of opaque and transparent strips of equal width. If the widths are unequal, then the resultant vector from each grating period is different and the intensity at the point S is different. However, the condition that the contribution from each period should add up in phase (i.e. interfere constructively) remains the same.

It is evident that with a series of opaque strips some of the light is merely absorbed by the grating. If, instead of absorbing the light corresponding to the return half of the vector diagram, we reverse its phase, then its contribution will add to, rather than cancel, the first half. This can be achieved by introducing a strip of transparent material with an *optical* thickness (i.e., physical thickness × refractive index) equal to half a wavelength. We then have the situation depicted in Figure 2.5(b) in which the amplitude is double that of the previous case and the irradiance at S is now I_0/π^2. So the grating has four times the efficiency of the first one.

Such a grating is known as a "phase grating" to distinguish it from the first type which is known as an "amplitude grating". In practice most spectroscopic gratings are both phase and amplitude gratings, although in general it is the phase aspect which is most important.

If we consider the vector diagram of Figure 2.4, we see that the total amplitude which is available is equal to the circumference of the circle, but, because of the variation of phase from contributions from different parts of the grating period, the maximum amplitude attained is equal to the diameter of the circle. If we could vary the phase across each individual groove, introducing an appropriate amount of optically dense material in such a way that each contribution was in phase, then the vector sum would simply become a straight line. The amplitude of the diffracted wave would be increased by a factor of π and the efficiency of the grating by π^2. Such a grating is shown in Figure 2.5(c) and is known as a blazed grating. The phase is constant from A to B′ where it jumps by 2π, and stays constant until C′ where it jumps again. If, instead of returning from C to C′, one continues the material in the line AB′ then this is equivalent to adding an extra retardation of 2π for the contribution for BC′, 4π for CD′, etc., so that each contribution is exactly in phase (as opposed to being out of phase by integral numbers of 2π). The reader may recognize this as a prism!

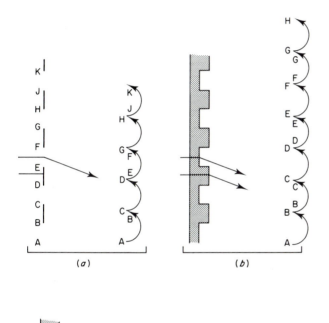

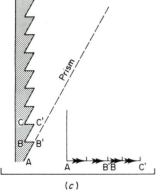

Figure 2.5 Various forms of diffraction grating.

☐   Complex amplitudes — multiple slits

We have so far considered the case of an aperture which was finite in one direction and infinite in the other and which was uniformly illuminated by a plane wavefront. In this case the vector sum was fairly straightforward, but in most cases this is not so and we need a more powerful technique to cope with the necessary mathematics. A concept which is very valuable in this context is that of *complex amplitudes*. Here, instead of representing the disturbance of the wave as a sine function we write it in an exponential form where e is the base of natural logarithms and $i = \sqrt{-1}$.

$$y = ae^{i2\pi[\nu t - (x/\lambda)]} = ae^{i2\pi\nu t} \cdot e^{-i2\pi x/\lambda}$$

In fact, $e^{i\phi} = \cos\phi + i\sin\phi$, for any angle ϕ, so that we have both a sine and a cosine function, but at the end of the calculation we can take either the real (cosine) or imaginary (sine) part of the result if required. This representation is equivalent to that of the amplitude vector described in the previous section. In most applications we can ignore the time-varying factor $e^{i2\pi\nu t}$ since for the moment we are limiting our discussion to waves of the same frequency and this factor will be the same throughout. It merely reminds us that we are dealing with a wave motion of frequency ν. The factor $a^{i2\pi x/\lambda}$ is known as the complex amplitude of the wave and may be represented as a vector on an "Argand diagram" in which the real component is plotted on the x axis and the imaginary along the y axis. This is shown in Figure 2.6(a).

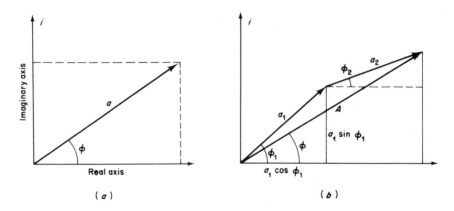

Figure 2.6 The representation of the complex amplitude on an Argand diagram.

If we wish to add the contributions of two or more waves it is sufficient simply to add their complex amplitudes and this is equivalent to the vector addition of their real amplitudes. Thus, in Figure 2.6(*b*),

$$Ae^{i\phi} = A(\cos \phi + i \sin \phi)$$
$$= a_1 \cos \phi_1 + a_2 \cos \phi_2 + i(a_1 \sin \phi_1 + a_2 \sin \phi_2)$$
$$= a_1 e^{i\phi_1} + a_2 e^{i\phi_2}$$

The resultant irradiance is given by the square of the real amplitude and this may be found by multiplying the complex amplitude with its complex conjugate. Thus we have

$$I = A^2 = Ae^{i\phi} \times Ae^{-i\phi} = A^2 e^{i(\phi - \phi)} = A^2$$

or

$$(A \cos \phi + iA \sin \phi)(A \cos \phi - iA \sin \phi)$$
$$= A^2(\cos^2 \phi - i^2 \sin^2 \phi) = A^2$$

We can consider a diffraction grating as a series of N equally spaced slits each contributing an amplitude a and differing in phase from its neighbour by an amount δ. The complex amplitude from the nth slit is then $ae^{in\delta}$ and the resultant from the whole grating is the sum

$$Ae^{i\phi} = a(1 + e^{i\delta} + e^{i2\delta} + e^{i3\delta} + \ldots e^{i(N-1)\delta})$$
$$= a \frac{1 - e^{iN\delta}}{1 - e^{i\delta}} \tag{2.6}$$

The irradiance is found by multiplying this expression by its complex conjugate, giving

$$A^2 = a^2 \frac{(1 - e^{iN\delta})(1 - e^{-iN\delta})}{(1 - e^{i\delta})(1 - e^{-i\delta})} = a^2 \frac{2 - 2\cos N\delta}{2 - 2\cos \delta}$$

which we may write as

$$A^2 = a^2 \frac{\sin^2 N\gamma}{\sin^2 \gamma} \tag{2.7}$$

where

$$\gamma = \frac{\delta}{2} = \frac{\pi d \sin \theta}{\lambda}$$

and a^2 is the irradiance of the diffraction pattern of each individual slit which may be written as $a^2 = (\sin^2 \beta)/\beta^2$, so that the complete description of the distribution of light from a grating consisting of a series of N slits is given by

$$I = A^2 = \frac{\sin^2\beta}{\beta^2} \cdot \frac{\sin^2 N\gamma}{\sin^2\gamma} \qquad (2.8)$$

Let us now consider the function $(\sin^2 N\gamma)/\sin^2\gamma$. We can see immediately that the numerator will be zero when $N = 0$, π, $2\pi, \ldots p\pi$, where p is an integer. The function will therefore vanish whenever $\gamma = p\pi/N$ except for $p = 0$ and $p = mN$ (m is an integer). In this case $\gamma = m\pi$ and the denominator also vanishes so the function becomes indeterminate. However, it may be shown that

$$\lim_{\gamma \to m\pi} \left(\frac{\sin N\gamma}{\sin\gamma} \right) = \lim_{\gamma \to m\pi} \frac{N \cos N\gamma}{\cos\gamma} = \pm N$$

In other words, whenever $\gamma = m\pi$ the function has a maximum value and the intensity is equal to N^2 times the intensity diffracted by a single slit. In between the minimum values corresponding to $\gamma = p\pi/N$ the function has a series of subsidiary maxima whenever $\gamma = (p\pi/N) + (\pi/2)$. Here the numerator has the value unity and the denominator has the value

$$\sin \pi \left(\frac{p}{N} + \frac{1}{2} \right)$$

The resulting distribution of intensity takes the form shown in Figure 2.7 and consists of a series of principal maxima corresponding to $\gamma = m\pi$ (i.e. $(\pi d \sin\theta)/\lambda = m\pi$ or $d \sin\theta = m\lambda$) and between these main maxima the function oscillates between zero and $N - 2$ secondary maxima which decrease in intensity the further they are from a principal maximum. The first minimum either side of a principal maximum occurs when $\gamma = \pm \pi/N$, so the angular half-width θ' of the principal maximum is given by

$$\gamma = \frac{\pi}{N} = \frac{\pi d \sin\theta'}{\lambda} \quad \text{i.e.} \quad \sin\theta' = \frac{\lambda}{dN} = \frac{\lambda}{W} \qquad (2.9)$$

where W is the width of the grating. This is the same as the angular half-width of the diffraction pattern corresponding to a single slit of width equal to that of the grating. Indeed, in the region close to the principal maxima the function

$$\frac{\sin N\gamma}{\sin\gamma} \quad \text{where} \quad \gamma = \frac{W}{N} \frac{\pi \sin\theta}{\lambda}$$

is very similar to the function

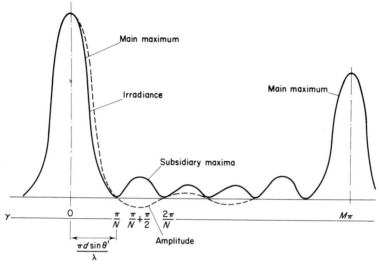

Figure 2.7 The diffraction pattern from multiple slits.

$$\frac{\sin \alpha}{\alpha} \quad \text{where} \quad \alpha = \frac{W\pi \sin \theta}{\lambda}$$

The greater the value of N the greater is the similarity.

☐   The general grating equation

In the previous section we considered the case of a plane wave impinging at normal incidence on a grating and found that the condition that light be diffracted towards the point S in Figure 2.3 was that $d = \lambda/\sin \theta$. Let us now consider the more general case when the light is incident at an arbitrary angle. This is shown in Figure 2.8 where for the sake of simplicity we represent each groove as a point. As we have seen, the magnitude of the contribution from each source will depend upon the form of the groove, but it will be equal for each groove. The condition that a diffracted order should exist is that the contributions from each groove should all be in phase, or rather out of phase by an integral number of 2π radians. This means that the difference in path from the source to the focussed image via successive grating grooves should be equal to a whole number of wavelengths. In other words, referring to the diagram

$$d \sin \alpha + d \sin \beta = m\lambda \qquad (2.10)$$

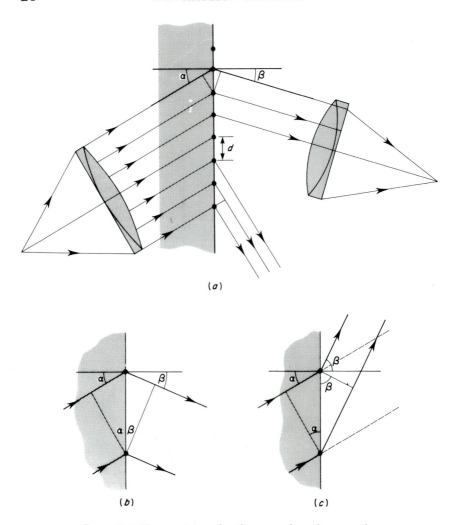

Figure 2.8 Nomenclature for the general grating equation.

where m is an integer known as the order number. This equation is of fundamental importance; it is known as "the grating equation" and will be the basis for many calculations of the properties of diffraction gratings. The integer m can be positive, negative or zero — zero obviously corresponds to the undeviated beam.

The precise form of the grating equation depends upon our choice of sign convention for the angles of incidence and diffraction. If we adopt the cartesian convention that angles in the first and third quadrant are positive and in the second and fourth quadrant are negative, then in Figure 2.8 β is negative and

$$d(\sin \alpha - \sin \beta) = m\lambda \qquad (2.10a)$$

On the other hand, if we adopt the convention that angles have the same sign when they are on the same side as the grating normal and opposite sign if the rays cross over the normal, then the equation is written as in (2.10). Both conventions ensure that for the un-deviated beam the signs of α and β are such that $m = 0$, and that an order which lies on the same side of the zero order as the grating normal has a positive value of m and one which is on the opposite side of the zero order from the normal has a negative value of m. The case of a negative order is shown in Figure 2.8(c) where either β is positive and greater than α and $d(\sin \alpha - \sin \beta) = m\lambda$, so $m < 0$, or β is negative because the ray has crossed over the normal and $d(\sin \alpha + \sin \beta) = m\lambda$, so m is negative. In practice the majority of gratings are used in reflection rather than in transmission, in which case in the diagrams of Figure 2.8 the diffracted beam is simply reflected about the plane of the grating as shown in Figure 2.9. If the cartesian sign convention is adopted then the sign of the angles must be reversed for the diffracted beam, because the light is now travelling from right to left. In the other convention the signs are unchanged.

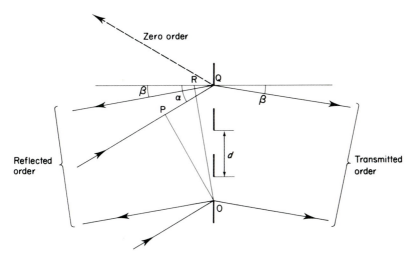

Figure 2.9 Sign convention for angles and order numbers.
Path difference $= PQ + QR$
In Convention (1): $d \sin \alpha + d \sin \beta = m\lambda$
In Convention (2): $d \sin \alpha - d \sin (-\beta) = m\lambda = d |\sin \alpha| + d |\sin \beta| = m\lambda$.

The grating equation (2.10) describes both the formation of the spectrum and the formation of the diffracted orders. Some insight into the way in which the various orders are formed may be gained by considering the vector sum in each case. This is shown in Figure 2.10 for the special case of an amplitude grating with equal bar-to-space ratio. In this case the vector sum goes round one complete revolution for the first order, two for the second, three for the third, etc. Since the second half is suppressed this gives rise to an amplitude of $1/2$ in the zero order, $1/2\pi$ in the first, 0 in the second (and all even orders) and $1/6\pi$ in the third (and $1/2m\pi$ in all odd orders). In this particular case the efficiency in the zero order is $1/4$, in the first order $1/4\pi^2$, in the third $1/36\pi^2$. However, we have seen in Figure 2.5 that if we control the variation of optical path across each individual groove we can straighten out the vector sum for any chosen order. In other words, by controlling the shape of the grooves we can blaze a grating for any order.

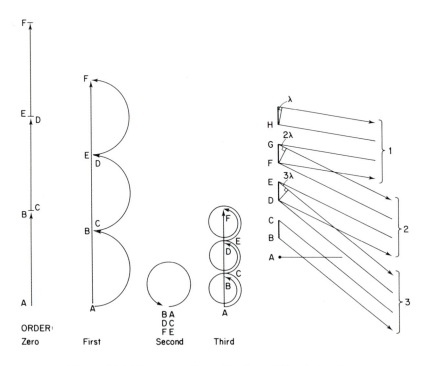

Figure 2.10 The generation of various diffracted orders.

☐ Dispersion

It is evident from the grating equation that the condition for the formation of a diffracted order depends upon the wavelength of the light. When we consider the formation of a spectrum we need to know how the angle of diffraction varies with wavelength and this we find very simply by differentiating the grating equation with respect to β.

$$\frac{\partial}{\partial \beta} \{d(\sin \alpha + \sin \beta) = m\lambda\}$$

$$d \cos \alpha \frac{\partial \alpha}{\partial \beta} + d \cos \beta = m \frac{\partial \lambda}{\partial \beta}$$

If we consider the case of a spectrograph where the angle of incidence is fixed then

$$\frac{\partial \alpha}{\partial \beta} = 0 \quad \text{and} \quad \frac{\partial \beta}{\partial \lambda} = \frac{m}{d \cos \beta} \qquad (2.11)$$

The quantity $\partial \beta / \partial \lambda$ is the change of diffraction angle corresponding to a small change of wavelength and is known as the *angular dispersion* of the grating. It is inversely proportional to $\cos \beta$, so that it increases with the angle of diffraction and tends to infinity as β tends to $90°$, that is, when the diffracted order grazes the surface of the grating. It is also a function of m/d, so that it is inversely proportional to the pitch of the grating and for a given pitch the dispersion is increased by using higher orders. Equation (2.11) gives the impression that the dispersion is independent of the angle of incidence. This is true provided that for a given grating at a given wavelength it is possible to find a value of α such that the required order is diffracted at the required value of β. It must be remembered that β is determined by the grating equation and as such is a function of m, d and α.

In practice what we require to know is not so much the angular dispersion but the linear dispersion in the plane of the focused spectrum. In other words, what change of wavelength corresponds to a given distance in the plane of the spectrum? The linear dispersion is simply the product of the angular dispersion and the focal length of the camera lens, i.e.

$$\frac{\partial x}{\partial \lambda} = f \frac{\partial \beta}{\partial \lambda}$$

provided that the focal plane of the instrument is perpendicular to the diffracted beam. Sometimes, however, in real instruments the focal plane is not normal to the diffracted beam but inclined at an angle ϕ. The linear dispersion must then be written as

$$\frac{\partial x}{\partial \lambda} = F f \frac{\partial \beta}{\partial \lambda}$$

where F is the so-called plate factor and is equal to $1/\cos \phi$. This occurs in practice when the camera lens has been designed to give a flat field (if the field were curved it would be necessary to bend photographic plates in order to keep them in focus) and in order to achieve this it is sometimes necessary to accept a design in which the focal length varies quite significantly with wavelength.

The dispersion considered so far refers to the situation in which light of all wavelengths is incident upon the grating at the same fixed angle as it is, for example, in a spectrograph where the spectrum is recorded on a photographic plate. Many gratings, however, are used in a monochromator in which light enters the instrument through a fixed slit and a section of the spectrum is selected by allowing it to pass through a second fixed slit. The spectrum is scanned by rotating the grating so that in the grating equation both α and β are varying with wavelength, with the constraint that $\alpha - \beta$ (or $\alpha + \beta$ depending upon the sign convention) remains constant. In such an instrument we often need to know the rotation of the grating that is required to give a particular shift in wavelength passing through the instrument. To do this we express α in terms of β and the angular deviation (i.e. the constant $\alpha - \beta = \delta$) and differentiate as before. A particularly simple case is that of the "Littrow mounting" in which the diffracted beam returns along the path of the incident beam. In this case $\alpha = \beta = \theta$ so the grating equation becomes[†]

$$2d \sin \theta = m\lambda$$

$$2d \cos \theta \, \partial\theta = m \, \partial\lambda$$

so

$$\frac{\partial \theta}{\partial \lambda} = \frac{m}{2d \cos \theta} = \frac{\tan \theta}{\lambda} \qquad (2.12)$$

In this case, the angle by which we must *rotate the grating* in order to achieve a given change in wavelength is just half the angle subtended by the same length of spectrum in a spectrograph. This

[†] The ideal Littrow mounting in which the diffracted beam returns exactly along the path of the incident beam is never used in practice. However, since under these circumstances the grating equation is greatly simplified, it is an interesting case to consider theoretically. In many cases it serves as an adequate approximation to configurations that are used in practice.

result is quite unsurprising since we are already familiar with the fact that if we rotate a mirror by an angle θ the reflected beam is rotated by 2θ. However, when working with a monochromator it is most important to bear in mind the distinction between the rotation of the grating and the angular spread of the spectrum.

☐   Resolution and resolving power

The terms "resolution" and "resolving power" are used to describe the ability to distinguish between adjacent components in a spectrum. The resolution of an instrument is the smallest change of wavelength that it enables us to detect. The resolving power of a grating is the ratio between the smallest change of wavelength that the grating can resolve and the wavelength at which it is operating.[†]

The resolving power is a property of the grating, whereas the resolution is a function both of the grating and the instrument in which it is put and may well be limited by other factors such as the resolution of the photographic plate or the size of the entrance (and/or exit) slit.

Let us consider the diffraction pattern in the focal plane of a spectrometer as shown in Figure 2.11.

The diffraction pattern of the aperture at the focal plane is represented by

$$I_x = I_0 \frac{\sin^2\gamma}{\gamma^2}$$

where γ is the difference in phase between contributions from the centre and the edge of the aperture. The first minimum occurs when this has a value of π, i.e.

$$\frac{W'}{2} \sin \partial\beta \frac{2\pi}{\lambda} = \frac{W'}{2} \frac{x_0}{f} 2\pi = \pi$$

so

$$x_0 = \frac{f\lambda}{W'}$$

[†] The term "resolving power" is not universally accepted. There are those who would argue that the use of the word "power" should be restricted to its mechanical meaning of the rate of doing work and therefore prefer to talk of the "resolution" of a grating. However, since this can lead to confusion between the "resolution" of an instrument (expressed in spectral units) and the "resolution" of a grating (which is dimensionless), we shall refer to the resolving power of a grating as a measure of its "ability to resolve".

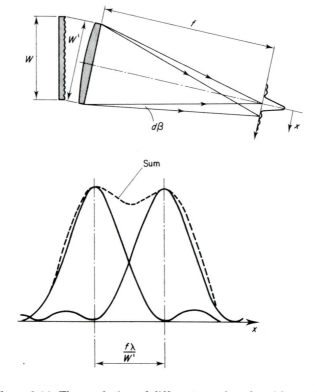

Figure 2.11 The resolution of different wavelengths with a grating.

If there are two wavelengths present and they differ by $\Delta\lambda$, then each will produce a diffraction pattern at the focal plane of the instrument, but according to Equation (2.11) they will be displaced by an amount

$$\Delta x = \frac{\Delta\lambda fm}{d \cos\beta}$$

In order to calculate the resolving power of the grating we must determine how close these two patterns can be without merging into each other. Rayleigh's criterion was that the two could still just be resolved if the maximum of one pattern coincided with the minimum of the other, i.e. $\Delta x = x_0$. So

$$\frac{\Delta\lambda fm}{d \cos\beta} = \frac{f\lambda}{W}$$

but since $W' = W \cos\beta$

$$\frac{\lambda}{\Delta\lambda} = \frac{Wm}{d} = mN$$

where N is the total number of grooves, or

$$\frac{\lambda}{\Delta\lambda} = \frac{W}{\lambda}(\sin\alpha + \sin\beta) = \frac{2W}{\lambda}\sin\theta \qquad (2.13)$$

in Littrow mounting.

Although the relationships expressed in (2.13) are almost universally used to express the resolving power of a perfect grating, it must be remembered that the Rayleigh criterion on which they are based is quite arbitrary. It assumes first that the images corresponding to the two wavelengths have the same intensity. If one diffraction pattern is far less intense than the other then it might give rise only to a slight asymmetry in the stronger pattern which would be very difficult to detect. On the other hand, the contrast of a photographic image can be enhanced chemically and there are other forms of image processing using computers to resolve a measured curve into its constituent components even though they may be impossible to discern with the naked eye. The form of the diffraction pattern will vary if the shape of the aperture is changed or if the distributions of light over the aperture is not uniform. These so-called "apodization" effects will also affect the resolution that can be achieved in practice. However, the Rayleigh criterion remains a very useful guide to resolution and is quite adequate for most purposes.

It is interesting to consider the physical significance of the expressions (2.13). The first tells us that for a given grating pitch of d used in a given order m the resolving power is directly proportional to the width of the grating. This is why we need large gratings for high-resolution work. It also tells us that it is possible to achieve the same resolution with a coarse grating used in a higher order as it is with a fine grating used in the first order. The second expression which is the simplest to remember tells us that the resolving power is the product of the order number and the total number of grooves on the grating. This is independent of the pitch of the grating (provided that the chosen order exists) but obviously the finer the pitch the smaller is the physical size of a grating containing a given number of lines.

The third expression involves neither pitch nor order number, but expresses the resolving power in terms of the angles of incidence and diffraction. It tells us that the resolving power is equal to the total number of wavelengths contained in the path difference between

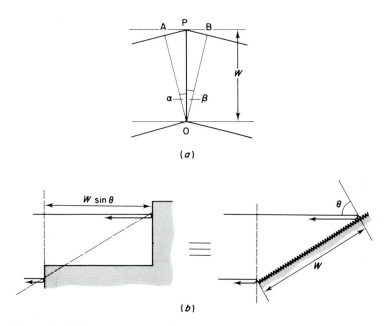

Figure 2.12 The path length difference introduced by various gratings.

light from the two edges of the grating. The total path difference as shown in Figure 2.12(a) is equal to

$$AP + PB = W \sin \alpha + W \sin \beta.$$

Therefore the resolving power is

$$RP = \frac{W(\sin \alpha + \sin \beta)}{\lambda}$$

or, in the Littrow case where the diffracted beam goes back along the direction of the incident beam, $\alpha = \beta = \theta$ and

$$RP = \frac{2W \sin \theta}{\lambda}$$

Thus in Figure 2.12(b) the resolving power of the two gratings shown is the same. The one consisting of two facets would be equivalent to a Michelson two-beam interferometer (or Young's slits) in which case the diffraction pattern is a simple sinusoidal function in which each peak of the sine wave corresponds to a diffracted order. Although the resolving power of such a device is the same as that of a grating, it has virtually no "free spectral range"

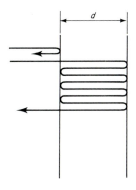

Figure 2.13 The path length difference for a Fabry–Perot interferometer.

(see following section) and would be of no practical use in that form. On the other hand, the diffraction grating could be regarded as a multiple-beam interferometer in which each groove contributes a "beam".

It is interesting to note that the physical significance of the resolving power is the same as for a multiple beam interferometer. For example consider the Fabry–Perot etalon in which the two plane parallel plates of reflectance r are separated by a distance d as in Figure 2.13. The resolving power is given by

$$ \mathrm{RP} \;=\; \frac{2d}{\lambda} \left\{ \frac{\pi r}{1 - r^2} \right\} $$

The expression $2d/\lambda$ is simply the number of wavelengths contained in the round trip between the mirrors; the expression $\pi r/(1 - r^2)$, the "finesse" of the etalon, can be interpreted as the effective number of inter-reflections that a beam makes before its contribution can be neglected. So here again the resolving power is equal to the number of wavelengths contained in the path difference between the extreme rays.

☐   Free spectral range and the overlapping of orders

One disadvantage of a grating compared with a prism is that it generates more than one diffracted order and therefore more than one spectrum. If a sufficiently wide range of wavelengths is incident upon the grating then these spectra will overlap and will be confused, as indicated in Figure 2.14. If the incident light contains all wavelengths from the infinitesimally small (say X-ray) to the very large (say microwaves or even radio waves) then there will always be confusion because at any wavelength λ for which the grating equation holds for an order m there will also be light (diffracted at the same angle) of wavelength $\lambda/2$ in order $2m$, $\lambda/3$ in order $3m$ and so on. The

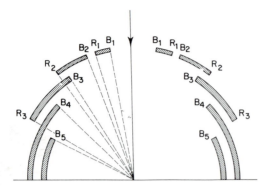

Figure 2.14 The overlapping of orders.

very long wavelengths are automatically taken care of by the grating equation itself because the pitch of the grating sets an upper limit to the wavelengths that will be diffracted. From the grating equation we have

$$\lambda = \frac{d}{m}\{\sin \alpha + \sin \beta\}$$

The maximum value of $(\sin \alpha + \sin \beta) = 2$ and the minimum value of $|m|$ is 1 for a useful order, so $\lambda_{max} = 2\lambda$ which corresponds to the Littrow mounting at grazing incidence and grazing angle of diffraction. At normal incidence $\sin \alpha = 0$ so that $\lambda_{max} = d$.

In many cases the problem of the diffraction of shorter wavelengths at higher orders is alleviated by other factors. For example, if the detector is incapable of detecting the complete range of wavelengths, then those outside its range will not matter. The first-order spectrum from any grating will be distinct when observed visually because the visible spectrum extends from about 400 nm to 700 nm which is less than a complete octave. The longest wavelength for which the second-order spectrum could overlap the first-order visible spectrum is 350 nm which is outside the range of human vision (although with a sufficiently strong source it may be detected due to fluorescence within the eye).

Air within the instrument will absorb all wavelengths below about 190 nm and often the grating efficiency falls off dramatically at short wavelengths depending upon the material of its surface. However, there are occasions, particularly in vacuum ultraviolet spectroscopy, when the overlapping of orders poses a very serious problem and it is necessary to introduce filters in order to disentangle the spectrum. The problem is nowhere more acute than in studies using synchrotron radiation, in which the source radiates a continuous spectrum from the X-ray region up to the infrared, and in many cases instruments have to be designed to operate over a substantial portion of this spectrum. Unfortunately, the spectroscopist is not even offered the alternative of using a prism because there do not exist suitable materials that will transmit radiation throughout the whole range.

The *free spectral range* of a grating is the largest bandwidth in a given order which does not overlap the same bandwidth in an adjacent order. Let the extremes of this band be represented by λ_1 and λ_2, where λ_2 is the longer wavelength, and let them be used in the order m. Overlapping will occur at the long-wave end of the band when λ_2 in order m is diffracted at the same angle as λ_1 in order $m + 1$ and will occur at the short-wave end when λ_1 in order m overlaps λ_2 in order $m - 1$. That is,

$$\lambda_2 m = \lambda_1(m + 1) \qquad \lambda_1 m = \lambda_2(m - 1) \qquad (2.14)$$

Hence, to avoid overlapping we require that

$$\lambda_2 - \lambda_1 \geqslant \lambda_1/m \quad \text{or} \quad \lambda_2 - \lambda_1 \geqslant \lambda_2/m \qquad (2.15)$$

However, since $\lambda_1 < \lambda_2$, the first expression is the more stringent we may say that the free spectral range $(\lambda_2 - \lambda_1)$ is equal to the shortest wavelength in the allowed bandwidth divided by the order number. This expression is perhaps the simplest form to remember, but it does lead us to assume that the longest wavelength is in fact diffracted. In the first order this is not necessarily the case, because it could happen that the longest wavelength in the spectrum is determined by the presence of the grating surface (i.e. $\beta = 90°$) rather than the overlapping of orders. This difficulty is overcome if we express the free spectral range in terms of the longest wavelength of the band, i.e.

$$\lambda_2 - \lambda_1 = \frac{\lambda_2}{m + 1} \qquad (2.16)$$

It is evident from Equations (2.15) and (2.16) that the free spectral range is reduced when the grating is used in higher orders. The problem is particularly important in the case of coarse gratings or "echelles" which are designed to work in high orders. For example, an echelle with 100 grooves mm^{-1} might be used in the 9th order at a wavelength of 500 nm, where it would have a free spectral range of only 50 nm. To overcome this problem echelles are used in conjunction with some other dispersing element, either a prism or a

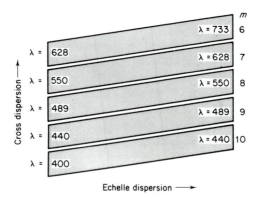

Figure 2.15 The form of the spectrum from an echelle.

grating which is so arranged that it disperses the spectrum in an orthogonal direction to the echelle spectrum. The spectrum would then be arranged in a raster form as shown in Figure 2.15.

The presentation of the spectrum in this form is a positive advantage, in many cases, because of its compactness. This is particularly so in astronomy where the light source is very weak and one is able to concentrate the whole of the spectrum onto the cathode of an image intensifier. Indeed, if the same spectrum at the same resolution were to be measured in first order it could subtend an angle exceeding 2π! (In other words in order to obtain sufficient resolution at short wavelengths in first order, the pitch of the grating would be such that the longest wavelength would be diffracted over the grating "horizon".)

☐   Calculation of facet angle for blazed gratings

The concept of blazing a grating is that each groove should be so formed that independently, by means of geometrical optics, it redirects the incident light in the direction of a chosen diffracted order. Thus in a transmission grating each groove constitutes a tiny prism and for a reflection grating each groove consists of a small mirror inclined at an appropriate angle. In practice the reflection case is more important, but we shall also consider transmission gratings if only to indicate some of the problems that are involved in their manufacture.

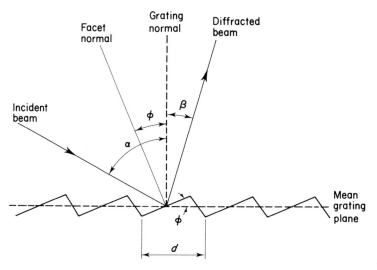

Figure 2.16 Determination of the facet angle for a blazed grating.

In Figure 2.16 the light is incident at angle α and diffracted at an angle β, where both are measured from the grating normal. The blaze condition is satisfied when the angle of incidence with respect to the facet normal is equal to the angle of reflection from the facet, i.e.

$$\alpha - \phi = \beta + \phi$$

or

$$\phi = \frac{\alpha - \beta}{2} \quad \text{or} \quad \beta = \alpha - 2\phi$$

which shows us immediately that the facet angle depends upon the mounting in which the grating is used. Thus, for example, if we consider the Littrow mount, $\alpha = -\beta$ so $\phi = \beta$, which is the situation depicted in Figure 2.17(a).

In this case the grating equation may be written

$$2d \sin \phi = m\lambda$$

so the facet angle is given by

$$\phi = \sin^{-1} \frac{m\lambda}{2d} \tag{2.17a}$$

and the planes defined by the facets are separated by m half-wavelengths.

In the case of normal incidence, as shown in Figure 2.17(b), we may set $\alpha = 0$ in the grating equation so that $\phi = -\beta/2$, but

$$d(\sin \alpha - \sin \beta) = m\lambda, \qquad \beta = \sin^{-1}(-m\lambda/d)$$

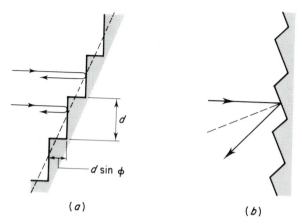

(*a*)   (*b*)

Figure 2.17 A blazed grating used (*a*) in the Littrow mounting, (*b*) at normal incidence.

so

$$\phi = \tfrac{1}{2} \sin^{-1}\left(\frac{m\lambda}{d}\right) \tag{2.17b}$$

which is to be distinguished from the previous expression.

Since, for a given wavelength, the facet angle varies according to the angle of incidence, it follows that a given grating will be blazed for different wavelengths when it is used at different angles of incidence. In general if α is the angle of incidence and ϕ is the facet angle, then the angle of diffraction β on blaze is equal to $\alpha - 2\phi$. So for the grating equation we may now write

$$d\{\sin\alpha - \sin(\alpha - 2\phi)\} = m\lambda_{\text{blaze}}$$

$$2d\sin\alpha\cos(\alpha - \phi) = m\lambda_{\text{blaze}}$$

but the blaze wavelength for the first-order Littrow configuration is given by

$$\lambda_{\text{b.Litt}} = 2d\sin\alpha$$

so we may write

$$m\lambda_{\text{blaze}} = \lambda_{\text{b.Litt}}\cos(\alpha - \phi) \tag{2.18}$$

This expression tells us that the blaze wavelength is equal to the Littrow blaze wavelength multiplied by the cosine of the angle of incidence on the facet, so that as the angle of incidence increases the blaze wavelength is reduced.

□   Transmission gratings

For the sake of simplicity we shall consider only the case of normal incidence. This enables us to ignore refraction at the back surface of the grating and to apply Snell's laws of refraction at the face of the miniature prism of the groove. If μ is the refractive index of the grating material then we may write

$$\mu \sin \phi = \sin (\phi + \beta)$$

so

$$\tan \phi = \frac{\sin \beta}{\mu - \cos \beta}$$

(see Figure 2.18). This can be resolved by determining β from the grating equation, which in this case can be written in the form $d \sin \beta = m\lambda$. So

$$\tan \phi = \frac{m\lambda}{d\mu - (d^2 - m^2\lambda^2)^{1/2}} \tag{2.19}$$

When performing these calculations it is important to realize that there is an upper limit to the value of ϕ and this is determined by total internal reflection at the facet. This occurs when the angle of refraction $(\phi + \beta)$ exceeds $90°$ in which case

$$\mu \sin \phi_{max} = 1 \qquad \phi_{max} = \sin^{-1}(1/\mu)$$

The exact value will naturally depend upon the material of the grating, but in most cases is about $40°$. For example, a 1200 groove mm^{-1} grating blazed at 500 nm would be impracticable if $\mu = 1.5$.

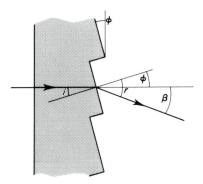

Figure 2.18 A blazed grating used in transmission.

The object of blazing a grating is to ensure a constant optical path in the diffracted beam across the area corresponding to a given facet. The phase lag introduced by a given depth of material h in a transmission grating is

$$2\pi h(\mu - 1)/\lambda$$

whereas in a reflection grating the optical path introduced by a groove of depth h is $2h$, so the phase lag is

$$(2\pi/\lambda)\,2h.$$

Thus, for a material of refractive index 1.5 we see that the grooves on a transmission grating need to be approximately four times the depth of those of the equivalent reflection grating.

Let us consider the case of a grating of 600 grooves mm^{-1} blazed for 500 nm in the first order at normal incidence and which is made of a material with a refractive index of 1.5.

In reflection we have:

$$\phi = \tfrac{1}{2}\sin^{-1}(m\lambda/d)$$
$$= \tfrac{1}{2}\sin^{-1}(0.5/1.67) = 8.7°$$

In transmission:

$$\beta = \sin^{-1}(m\lambda/d) = \sin^{-1}0.3$$
$$\tan\phi = 0.3/(1.5 - 0.95)$$
$$\phi = 28.6°$$

In fact in the small angle approximation where $\beta = \sin\beta = \tan\beta$ and $\cos\beta = 1$:

$$\phi_{\text{refl}} = \frac{m\lambda}{2d}$$

$$\phi_{\text{trans}} = \frac{m\lambda}{d}\left(\frac{1}{1.5-1}\right) = \frac{2m\lambda}{d} = 4\phi_{\text{refl}}$$

☐   Limitations of the theory of blazing

The geometrical or scalar theory of gratings which has been presented so far provides us with a good insight into the way in which a grating works. It enables us to derive accurately most of the equations that are required and it enables us to calculate most of the features of a grating that are needed for the design of an instrument. However, the scalar theory is not exact and its shortcomings are particularly apparent when we come to consider efficiency characteristics. In order to describe fully the operation of a grating we must consider the interaction of the incident electromagnetic wave with the material of the grating and for this we must solve Maxwell's equations for the periodic boundary defined by the grating surface.

The scalar theory, based on the Huygens principle of secondary wavelets, assumes that the optical properties of the material of the grooves are the same on a microscopic scale as they are on a macroscopic scale. In other words, each groove facet is a small mirror or a small prism but behaves in the same way as a large one. When the widths of the grooves are comparable with the wavelength of light this assumption is no longer valid. For example, consider the operation of a metal mirror. The incident electromagnetic wave induces oscillations of the free conduction electrons in the surface of the metal. These behave as a series of oscillating dipoles which re-radiate in phase with the incident wave and may be considered as a series of secondary sources which interfere constructively in the direction of the reflected wave in accordance with Huygens's theory. But when the dimensions of the facet are of the same order as the wavelength of the light, then the oscillations of the electrons are impeded or curtailed by the boundary of the facet and the simple scalar theory no longer applies.

The breakdown of the scalar theory first manifests itself with the appearance of polarization effects. The simple theory applies equally to light of either polarization and, even if it is modified to include the macroscopic effects of polarization upon reflection away from normal incidence, the results are only slightly changed. In practice when the groove spacing is less than about three times the wavelength the efficiency curves for the two polarizations differ dramatically and the wavelengths of peak efficiency, the "blaze wavelengths", differ. In general the grating approximates best to the scalar theory for light polarized with its electric vector parallel to the grooves and deviates most when the electric vector is across the grooves. This is consistent with our description of the oscillation of conduction electrons in the surface. If they oscillate along the grooves, they are not impeded by the finite length of their path although they may be perturbed by the presence of electrons oscillating in the adjacent grooves. If they oscillate across the grooves, then their path is curtailed by the edge of the groove and the net electric field is determined largely by edge effects which may safely be ignored when considering reflections from a large mirror. The consequences of the breakdown of scalar theory are considered in more detail in Chapter 6. In general only the efficiency of the grating is affected and provided that this is known empirically it does not usually affect the design of grating instruments. The scalar theory enables us to understand a great deal about gratings and is far more convenient than the full electromagnetic theory. However, its shortcomings should always be borne in mind.

□   A description of a grating in terms of the
           Fourier transform

The expression for the sum in the image plane of the components of amplitude due to contributions from the whole of an aperture as shown in Figure 2.3 may be written in an integral form in terms of complex amplitudes:

$$G(s) = \int_{-\infty}^{\infty} \rho(x) \exp \{2\pi i s x\} \, dx$$

where $G(s)$ is the resultant amplitude at a point s in the plane of the focused diffraction pattern defined by $(\sin \theta / \lambda) = s$ and $\rho(x)$ is the function describing the amplitude of the wavefront at a position x in the plane of the aperture. This operation of multiplying a function $\rho(x)$ by an appropriate phase factor and integrating exists in mathematics in its own right and is known as a Fourier transformation. The function $G(s)$ is known as the Fourier transform of $\rho(x)$. In optical terms, $G(s)$ is simply the Fraunhofer diffraction pattern of the object represented by $\rho(x)$. It is beyond the scope of this book to describe the theory of Fourier transformation, but we shall see that by using some of the results of this theory we can arrive at a very succinct description of the diffraction grating that is particularly useful in providing an insight into the way in which any given feature of a grating will affect its performance.

We shall start by stating some of the basic concepts and fundamental relationships concerning the Fourier transform. We shall not prove their validity, but where appropriate we shall draw attention to their optical analogue and invite the reader to accept their plausibility.

The Dirac delta function $z(x)$

This is a function in which the amplitude density is zero everywhere except at one point x where the amplitude is infinite:

$$\rho(x) = \infty \qquad x = x_0$$
$$\rho(x) = 0 \qquad x \neq x_0$$

The integrated amplitude is equal to

$$\int_{-\infty}^{\infty} \rho(x)\, dx = \infty \times 0$$

since the width is infinitesimal. This is indeterminate, but is defined to be equal to unity:

$$\int_{-\infty}^{\infty} z(x)\, dx = 1$$

This function is the mathematical equivalent of a point source. Consider it as a pinhole which transmits unit amplitude but is of zero (or in practical terms, negligible) dimensions. The Fourier transform of a delta function at a position x is given by

$$G(s) = \int_{-\infty}^{\infty} z(x) \exp\{2\pi i s x\}\, dx = \exp\{2\pi i s x\}$$

This function has a constant amplitude of unity, but has associated with it a phase factor $\exp\{2\pi i s x\}$ which varies with x. In other words, with an ideal point source the intensity which is observed is constant but its phase depends upon its position.

Let us now consider the Fourier transform of a pair of delta functions symmetrically displaced about the origin at a distance $a/2$ as shown in Figure 2.19.

$$G(s) = \int_{-\infty}^{\infty} \rho(x) \exp\{2\pi i s x\}\, dx$$
$$= \exp\left\{2\pi i s \frac{a}{2}\right\} + \exp\left\{2\pi i s \frac{-a}{2}\right\}$$
$$= 2 \cos(\pi s a)$$

Object

Fourier transform

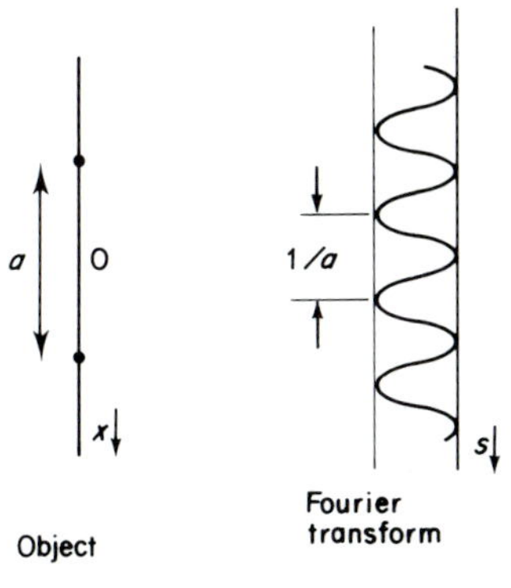

Figure 2.19 The Fourier transform of a pair of delta functions.

This describes the fringe pattern which is the familiar result of the experiments of Young's slits. It also illustrates one very important feature of the Fourier transform, namely that the dimensions of the Fourier transform are inversely proportional to the dimensions of the subject. Thus, if we double the separation of Young's slits we halve the distance between consecutive fringes. In general the Fourier transform will generate for any object in space a corresponding object with reciprocal dimensions and will define a so called "reciprocal" space. This concept is of great importance in crystallography when one is dealing with objects in three dimensions such as a crystal lattice. However, in physical optics one is generally concerned with two-dimensional objects and often when considering diffraction gratings we assume that the properties are constant along the length of the groove and restrict ourselves to a one-dimensional object. (We have in fact already made this assumption in the analogy between two *point* sources and Young's *slits*.)

A second interesting feature of the Fourier transform is that the Fourier transform of a Fourier transform is equal to the original function. In other words, if

$$G(s) = \int_{-\infty}^{\infty} \rho(x) \exp\{2\pi isx\} \, dx$$

then

$$\int_{-\infty}^{\infty} G(s) \exp\{2\pi isx\} \, ds = \rho(x)$$

Consider for example the case of Young's slits. The first transform generates a sinusoidal amplitude distribution such as that transmitted by a sinusoidal amplitude grating. In the ideal case such a grating generates only two diffracted orders $m = +1$ and $m = -1$ which in Fraunhofer diffraction (i.e. in the second transform) will be brought to a focus at two lines which are the equivalent of the slits that we started with.

A second function which is particularly relevant in optics is the so-called "top-hat" function in which $\rho(x) =$ unity between some limits and zero everywhere else. This represents a finite, uniformly illuminated aperture. Let it have a total width a and let it be symmetrically positioned with respect to the origin as shown in Figure 2.20.

$$G(s) = \int_{-\infty}^{\infty} \rho(x) \exp(2\pi isx) \, dx$$

$$= \int_{-a/2}^{a/2} \exp(2\pi isx) \, dx = \left[\frac{\exp(2\pi isx)}{2\pi is} \right]_{-a/2}^{a/2}$$

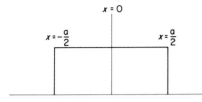

Figure 2.20 The "top-hat" function representing a uniformly illuminated aperture.

$$= \frac{\cos(\pi as) + i\sin(\pi as) + \cos(-\pi as) - i\sin(-\pi as)}{2(\pi is)}$$

$$= \frac{\sin(\pi as)}{\pi s} = \frac{a\sin\beta}{\beta}$$

where $\beta = \pi as$ and is the phase difference between the edge and the centre of the aperture. This is the familiar expression for the amplitude of the diffraction pattern of a slit of finite width.

If we consider the case of two top-hat functions of width a separated by a distance b (i.e. two slits of finite width) (Figure 2.21) we obtain another familiar result and one which illustrates another important feature of the Fourier transform.

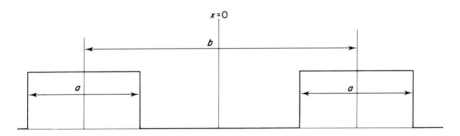

Figure 2.21.

The aperture function is given by

$$\rho(x) = 1 \qquad -\left(\frac{a+b}{2}\right) < x < \left(\frac{a-b}{2}\right)$$

$$\rho(x) = 1 \qquad \left(\frac{b-a}{2}\right) < x < \left(\frac{a+b}{2}\right)$$

$$\rho(x) = 0 \qquad \text{elsewhere}$$

We may therefore write the Fourier transform as

$$G(s) = \int_{-(a+b)/2}^{(a-b)/2} \exp(2\pi i s x)\, dx + \int_{(b-a)/2}^{(a+b)/2} \exp(2\pi i s x)\, dx$$

$$= \left[\frac{\exp(2\pi i s x)}{2\pi i s}\right]_{-(a+b)/2}^{(a-b)/2} + \left[\frac{\exp(2\pi i s x)}{2\pi i s}\right]_{(b-a)/2}^{(a+b)/2}$$

$$= 2\cos\pi b s \cdot \frac{a\sin\beta}{\beta}$$

(if we again substitute $\beta = a\pi s$) which is simply the product of the Fourier transform of two infinitesimal slits and the Fourier transform of the single finite slit. This result is quite general in that whenever two functions are *convoluted* the Fourier transform of the result is the product of the Fourier transform of the individual functions. Convolution is the action of repeating one function at every point of another function and is represented by the symbol $\otimes$. Thus Figure 2.22(a) shows the convolution, and Figure 2.22(b) the result when Fourier transformed. It may also be shown that the converse is true. The Fourier transform of the product of two functions is the convolution of their individual Fourier trasnforms. To summarize:

The Fourier transform is a mathematical operation which relates functions in real space to a corresponding set of functions in "reciprocal space". An example of this is the relationship between an object and its Fraunhofer diffraction pattern.
Features which are large in real space are small in reciprocal space and vice versa.
A multiplication in real space leads to a convolution in reciprocal space.
A convolution in real space leads to a multiplication in reciprocal space.
A shift in real space leads to a change in phase in reciprocal space.

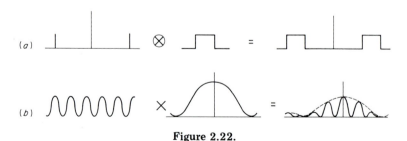

Figure 2.22.

Let us now consider a grating within this framework. Our starting point is a "Dirac comb": a function which consists of an infinite series of delta functions separated by a distance d (Figure 2.23).

Figure 2.23 The "Dirac comb".

The Fourier transform is simply

$$\sum_{n=-\infty}^{n=0} \exp{(2\pi inds)}$$

or, if we take the terms in pairs corresponding to $+n$ and $-n$,

$$\sum_{n=0}^{n=\infty} \exp{(2\pi inds)} + \exp{(-2\pi inds)} = \sum_{n=0}^{n=\infty} 2\cos{(2\pi nds)}$$

If ds is an integer $= m$ then the value of $\cos{(2\pi nds)}$ is unity. If ds is not an integer then the series sums to zero because for any value of $(2\pi nds)$ later in the series there will be a corresponding value of $(2\pi nds + \pi)$ which will cancel it.

The Fourier transform of a Dirac comb of spacing d is itself a Dirac comb but of spacing $1/d$ since the function is zero everywhere except when $s = m/d$. If we now remind ourselves that s is defined as $(\sin\theta/\lambda)$ then we have

$$\frac{\sin\theta}{\lambda} = \frac{m}{d} \quad \text{or} \quad d\sin\theta = m\lambda$$

which is of course the grating equation for normal incidence and the peaks of the Dirac comb in the Fourier transform are simply the diffracted orders of the grating.

In practice gratings are not infinite in extent and we may represent this by multiplying the Dirac comb by a top-hat function, as shown in Figure 24(a). The result of this is that in reciprocal space the Dirac comb is convoluted with the sinc function $(\sin\beta)/\beta$ which is the Fourier transform of the finite aperture, and this is shown in Figure 24(b).

Thus we see that in practice the spectral image has a finite width which is inversely proportional to the width of the grating and that

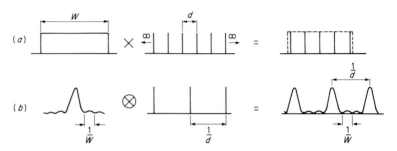

Figure 2.24

therefore the resolving power of the grating is proportional to the width. It also shows that there are

$$\frac{1/d}{1/W} - 2 = \frac{W}{d} - 2 = N - 2$$

secondary maxima (where N is the total number of grooves) between the main diffracted orders.

Again, in practice, gratings do not consist of series of infinitely narrow slits but of grooves of finite width and of recognizable profile. The groove profile is then convoluted with the basic Dirac comb and we would therefore expect that the reciprocal Dirac comb would be multiplied by the Fourier transform of the groove profile so that the various spikes would have different intensities. This is simply another way of saying that the form of the groove determines the distribution of light among the diffracted orders. However, it is at this stage that we must bear in mind the limitations of Huygens's principle. When the spacing of the grooves is of the order of the wavelength of light the envelope function multiplying the diffraction pattern is not necessarily the Fourier transform of the groove profile.

With that reservation in mind we may now regard a diffraction grating as a Dirac comb multiplied by a top-hat function and convoluted with a groove profile. In the plane of the focused spectrum we observe, for each wavelength incident upon the gratings, its Fourier transform which consists of a Dirac comb (the orders of diffraction) convoluted with a sinc function (due to the finite width) and multiplied by an envelope function (the blaze effect of the grooves). In real space we have Figure 2.25(a) and in reciprocal space, Figure 2.25(b).

This description gives us a neat summary of the properties of a grating and demonstrates the way in which the various aspects of a grating affect its performance. The results relating to a perfect

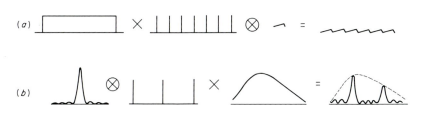

Figure 2.25.

grating have already been obtained by other means, but the value of the Fourier transform view of a grating is that it gives us some immediate intuitive insight into the effects of deviations from perfection.

Suppose, for example, that the groove profile, instead of being a triangle was not ruled sufficiently deeply and there was a flat "land" in between the grooves. Clearly this will affect the envelope function and hence the distribution of light among the orders, but it will have no effect on the resolution or the position of the diffracted orders. On the other hand, if the grooves are rough and have jagged and irregular edges, then we can consider this as the product of the ideal grating with an aperture function containing all of the information about the roughness. The roughness will manifest itself as a random variation in phase and amplitude across the aperture and this will generate diffuse scattering. The Fourier transform of the aperture function is no longer a sinc function but one in which energy has been transferred from the peak of the curve into the wings. The width of the curve is practically unaltered, so we see that the net effect on the grating is that the level of stray light is increased, resolving power is more or less unchanged, and the efficiency is reduced slightly, but the distribution of light among the diffracted orders is much the same (Figure 2.26).

Probably the most important example of a grating imperfection is that due to periodic errors in the grooves. If a groove is in the wrong position, then there will be a change of phase in the contribution from that groove; if its depth is different, then there will also be a corresponding change in amplitude. Again we can separate the effects of the perfect grating from its imperfections and simply regard the problem as that of an aperture function having periodic variations of phase and amplitude. Instead of the top-hat function we now have something that itself resembles a grating (Figure 2.27(a)). We know that the Fourier transform of such a

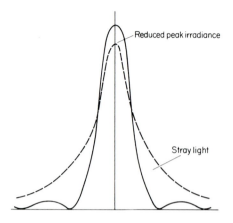

Figure 2.26 The effect of roughness on the spectral image of a grating.

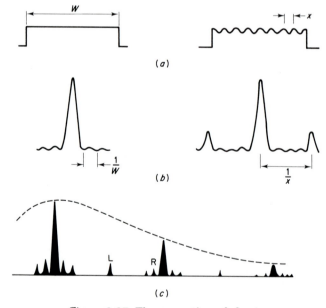

Figure 2.27 The generation of ghosts.

function is a series of diffracted orders separated by a distance $1/x$ (Figure 2.27(b)). This is the function which we must now convolute with the Dirac comb corresponding to the main grating, and gives us the (unfortunately) familiar result of the diffracted orders flanked by ghosts (Figure 2.27(c)). There are Rowland ghosts (R) close in to the diffracted order that are due to long-term periodic errors on the grating and Lyman ghosts (L) that are widely separated from the parent line and are due to short-term periodic errors in groove position (e.g. every two or three grooves).

A further example of the way in which the Fourier transform approach provides a direct and easily visualized solution is the question of a grating that is not uniformly illuminated. Let us suppose that, in order to ensure that all available light passes through the instrument, the grating has been underfilled. Let us also assume, as may often be the case, that the distribution of the light is gaussian. We may therefore multiply the top-hat aperture function by a gaussian function of the appropriate width. In fact, if the amplitude at the edges of the grating is very small, we may neglect the effect of truncation and simply replace the top-hat function by the gaussian one. The Fourier transform of a gaussian function is itself gaussian (see references), so the effect in the spectral plane of the grating is to replace the oscillatory sinc function by a gaussian. This will generate a slightly broader image as shown in Figure 2.28, but it will not have the secondary maxima that are associated with the sinc function. In fact the energy which would have gone into the secondary maxima has been used in broadening the main peak, so with gaussian illumination the resolving power of the grating is somewhat reduced, but the level of light to be found between the diffraction orders is also significantly reduced.

Figure 2.28 The effect of apodization by a gaussian amplitude function.

☐   A quantum mechanical description of gratings

The operation of a diffraction grating is traditionally described in terms of the wave nature of light. Indeed, we have already described a grating as an extension of Young's slits and it was Young's experiment which first provided convincing evidence for the wave nature of light. However, we now accept that all matter behaves both as particles and as waves, the properties of which are related by the de Broglie relationship

$$\lambda = \frac{h}{p}$$

where λ is the wavelength associated with a particle, h is Planck's constant and p is the momentum of the particle.

The "particle" of light is known as a photon and if we invert this relation we see that the momentum associated with a photon is given by

$$p = \frac{h}{\lambda}$$

The energy of a photon is given by Planck's empirical result

$$E\,(\text{energy}) = h\nu$$

where ν is the frequency of the photon ($= c/\lambda$).

We must remember that momentum is a vector quantity and as such it is specified by both its magnitude and its direction. Consider the reflection of a photon by a mirror. We may represent the incident photon and the reflected photon by vectors of length p in the directions of incidence and reflection. (Since $p \propto 1/\lambda$ this diagram is in fact drawn in reciprocal space.) The momentum of the incident photon can be resolved into two components, $p \cos \alpha$ normal to the mirror surface and $p \sin \alpha$ parallel to the mirror. Upon reflection the component perpendicular to the mirror is reversed in direction and the parallel component is unaltered (Figure 2.29).

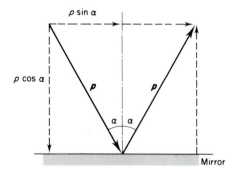

Figure 2.29 The effect of reflection of the momentum of a photon.

The action of a diffraction grating is to redirect the incident light into the different orders and in so doing it imparts momentum to the photon. Let us suppose, on the grounds of plausibility, that by analogy with the de Broglie relationship the momentum associated with a grating is h/d, and that this is in the plane of the grating and is normal to the grooves. The function of a grating then is to add quanta of momentum of this value to the in-plane component of the momentum of the photon. Upon diffraction the parallel component of the photon momentum becomes in general

$$p \sin \alpha + \frac{mh}{d}$$

However, since the wavelength does not change upon diffraction, neither does the magnitude of the momentum. The vector representing the diffracted photon momentum must be on a circle of radius $|p|$. It is then evident from Figure 2.30 that if β is the angle of diffraction, then

$$p \sin \beta = p \sin \alpha + \frac{mh}{d}$$

Substituting for p we have

$$\frac{h}{\lambda} \sin \beta = \frac{h}{\lambda} \sin \alpha + \frac{mh}{d}$$

which is the same as the grating equation when we take into account the appropriate sign convention for α, β and m.

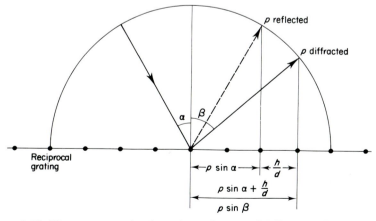

Figure 2.30 The representation in reciprocal space of diffraction from a grating.

This demonstration should not be regarded as an alternative derivation of the grating equation. It is based on plausibility guided by the knowledge from wave optics of what the final result must be. However, its inclusion here serves to remind us that there does exist a quantum mechanical description of the diffraction grating. We shall see later that there are occasions when the incident light interacts with the material of the grating and that it is then the quantum mechanical approach that gives the most succinct description of the phenomenon.

A further interesting application of the results of quantum mechanics to grating theory concerns the resolving power of the grating (Dravins 1978). In an experiment to measure simultaneously the position and the momentum (or wavelength) of a photon, the limits of uncertainty in these quantities are related by the Heisenberg uncertainty principle. This states that

$$\Delta x \, \Delta p \simeq h$$

where Δx is the uncertainty in position, Δp is the uncertainty in momentum and h is Planck's constant (see Figure 2.31).

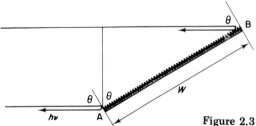

Figure 2.31 The uncertainty of the position of a diffracted photon.

In the case of a photon which has been diffracted by a grating in the Littrow configuration, the uncertainty in position is equal to $2W \sin \theta$. We cannot say whether the photon was diffracted from point A or point B or any other intermediate point. From de Broglie we know that the momentum of a photon is given by

$$p = \frac{h}{\lambda}$$

By differentiation:

$$\Delta p = -\frac{h}{\lambda^2} \Delta \lambda$$

In this case the negative sign has no significance so we may write

$$(2W \sin \theta)\left(\frac{h}{\lambda^2} \Delta\lambda\right) = h$$

or

$$\frac{\Delta\lambda}{\lambda} = \frac{2W \sin \theta}{\lambda}$$

i.e. the resolving power is equal to the number of wavelengths contained in the difference in optical paths from the two extremes of the grating. This is, of course, the same result as we obtained in Equation (2.13).

Bibliography

The basic theory of diffraction gratings is described in most standard textbooks on optics, such as:

Born, M. and Wolf, E. (1965). "Principles of Optics". Pergamon Press, Oxford.
Jenkins, F. A. and White, H. E. (1957). "Fundamentals of Optics". McGraw-Hill, New York.

and also in some detail by Stroke (1967), pp. 444–461 and 514–524.[†]

For a fuller treatment of Fourier transforms see:

Champeney, D. C. (1973). "Fourier Transforms and their Physical Applications". Academic Press, London and New York.

and

Lipson, S. G. and Lipson, H. (1969). "Optical Physics". Cambridge University Press.

[†] See main references.

3

The Use of Gratings in Spectroscopic Instruments

z

□   Spectroscopic instruments

The purpose of this chapter is to consider some of the ways in which gratings are used in spectroscopic instruments. A full study of the design of such instruments is beyond the scope of this book, but it is important to look briefly at some of the principles involved. The requirements of an instrument determine the parameters of the ideal grating for that application and often the performance of the instrument is restricted by the limitation of the techniques for making gratings. The performance of a grating can only be judged adequately in the context of the instrument for which it is intended and the art of designing an instrument is to achieve the best possible performance with the components that are available.

The basic elements of a spectroscopic instrument are shown in Figure 3.1. The source, or more usually an image of the source, fills an entrance slit and the radiation from this is collimated by either a lens or a mirror. The radiation is then dispersed, in our case by a grating, so that the direction of propagation of the radiation depends upon its wavelength. It is then brought to a focus by a second lens or mirror and the spectrum consists of a series of monochromatic images of the entrance slit. The focused radiation is detected, either by an image detector such as a photographic plate, or by a flux detector such as a photomultiplier, in which case the area over which the flux is detected is limited by an exit slit. In some cases the radiation is not detected at this stage, but passes through the exit slit to be used in some other optical system. As the exit slit behaves

z

57

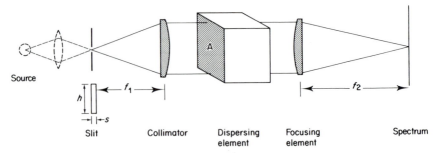

Figure 3.1 The basic elements of a spectroscopic instrument.

as a monochromatic source, the instrument can be regarded as a wavelength filter and is then referred to as a monochromator.

When the instrument is used with an image detector it is referred to as a spectrograph, except when the detector is the eye in which case the instrument is called a spectroscope. However, the terminology referring to the other mode of operation, that is with an exit slit, is not universally agreed. The term spectro*meter* is often used to describe an instrument where the spectrum is *measured* with a flux detector. One could argue, on the other hand, that a spectrograph also measures the spectrum and could also legitimately be described as a spectrometer. Equally it could be argued that whenever an exit slit is used, even if it is followed immediately by a flux detector, then the instrument is selecting a single wavelength band and could be called a monochromator. This would leave the term spectrometer free to describe all classes of instruments with which one can measure any feature of a spectrum. In this book we shall adopt (perhaps with some reluctance) the following terminology which appears to be that most widely used:

Spectroscopic instrument: any instrument capable of producing a spectrum.

Spectrograph: an instrument in which the image of the whole spectrum is detected on an image detector.

Spectroscope: an instrument in which the spectrum is viewed with the eye.

Spectrometer: an instrument in which the flux in a spectral image passes through an exit slit to be detected on a flux detector, and in which the spectrum is scanned by varying the wavelength which falls upon the exit slit.

Monochromator: an instrument which is used to select a given narrow band of spectrum by allowing only that portion of the spectrum to pass through an exit slit.

There are two fundamental differences between the operation of a spectrograph and that of a spectrometer. In a spectrograph the detector responds to the irradiance of radiation in the spectral image plane, that is to the flux density, whereas in a spectrometer it is the total flux which is measured. Therefore in a spectrograph it is advantageous that a given flux be brought to as small an image as possible, whereas in a spectrometer, provided that it does not overfill the detector, the size of the image is usually unimportant.[†]

The second point is that in general a spectrometer is significantly less efficient in collecting data than is a spectrograph, although it presents it in a far more convenient and quantified form. Consider for example a spectrum extending over a range of 300 nm which we wish to measure with a resolution of 0.1 nm. In a spectrograph all the radiation that enters the instrument and is not subsequently lost at the optical components is used in the recording of the spectral image. In a spectrometer at any instant only 1/3000 of the radiation forming the spectral image is allowed to pass through the slit and the rest is discarded. In applications where there is no shortage of light and where conditions are stable, so that it does not matter if it takes a long time to scan a spectrum, the convenience of obtaining the spectrum directly in a graphical form justifies this waste of energy. On the other hand, when the spectrum must be recorded as quickly as possible a spectrograph is preferred. In practice, however, the difference in efficiency is not always quite as great as this suggests, first because it is often practical to work at higher dispersion with a spectrometer and second because the quantum efficiency of a photographic plate is much lower than that of a photomultiplier.

Let us now consider in more detail the various components of a spectroscopic instrument. Suppose that the source has a radiance of $B\,\mathrm{W\,m^{-2}}$ per steradian. The total amount of radiant flux entering the instrument is

$$Bsh\,\frac{A}{f_1^2} \qquad (3.1)$$

[†] The truth of this statement depends upon the nature of the limiting noise in the system. At low light levels the optimum signal-to-noise ratio is obtained with as small a photosensitive area as possible.

Figure 3.2   The resolution of the spectral images of two wavelengths.

where s and h are the width and height of the entrance slit, f_1 is the focal length of the collimator and A is the area of the collimated beam intercepted by the dispersing element. (For a grating of aperture A', where the light is incident at an angle α, A is equal to $A' \cos \alpha$.)

Let us now assume that the resolution that is required is $\delta\lambda$ and that the angular dispersion of the grating is $d\beta/d\lambda$. If we assume that the widths of the slits are such that the effects of diffraction may be neglected, then two adjacent wavelengths are just resolved when their respective images of the entrance slit just touch, as shown in Figure 3.2.

We may therefore write

$$\delta\lambda = \frac{s'}{f_2} \frac{1}{d\beta/d\lambda} \tag{3.2}$$

where s' is the width of the monochromatic image of the entrance slit and f_2 is the focal length of the camera. The width of the entrance slit will be related to that of the exit slit by the magnification of the instrument in the plane of dispersion. So

$$s' = s \frac{f_2}{f_1} \frac{\cos \alpha}{\cos \beta}$$

The term $\cos \alpha/\cos \beta$ describes the anamorphism of the grating and arises because the width of the incident and diffracted beams are generally different. However, we may, without restricting the generality of our argument, consider the Littrow mounting in which $\alpha = \beta = \theta$ and $\cos \alpha/\cos \beta = 1$. We may therefore substitute in Equation (3.2) to obtain the relationship:

$$s = f_1 \frac{d\theta}{d\lambda} \delta\lambda$$

which can in turn be substituted in (3.1). If the coefficient of transmission of the optics (taking account of reflection losses at the collimator and camera and the efficiency of the grating) is τ_λ, the amount of flux Φ in the spectral image is

$$\Phi = B \frac{d\theta}{d\lambda} \delta\lambda \frac{hA}{f_1} \tau_\lambda \tag{3.3}$$

One might expect from this that the flux could be increased without limit simply by increasing the length of the entrance slit or by reducing the focal length of the collimator. In practice this is not so, because the quality of the image is limited by aberrations of the collimating and focusing optics. As the "field angle" of the system is increased, so the ends of the image of the slit become blurred and resolution is lost. Let $\beta = h/f_1$ (= the tangent of the field angle) be the maximum value for which an adequate image is achieved. In practice it is of the order of $1/100$ although it can be greater. The total flux in the image may then be written as

$$\Phi = B \frac{d\theta}{d\lambda} A \tau_\lambda \beta \delta\lambda \tag{3.3a}$$

In a spectrograph it is the flux density or irradiance rather than the total flux which is important, so we must develop this expression a little further. Given that there is a certain amount of flux Φ in the spectral image, then the density of flux is $\Phi/s'h'$, so that by reducing the image size the intensity will increase. We can only take advantage of this provided that the detector will still resolve the spectral image. In a photographic plate, for example, this limit is set by the grain size. Let us assume that the smallest practical image has a width $s_1 = g$. The width of the entrance slit is then given by $s = (f_1/f_2)g$[†] and this subtends an angle $\delta\theta = s/f_1$ at the grating.

As before, the spectral bandwidth corresponding to this angle is given by

$$\delta\lambda = \frac{d\lambda}{d\theta} \delta\theta$$

so, in order to resolve a given value of $\delta\lambda$, the width of the entrance slit is given by

[†] Neglecting any anamorphism.

$$s = f_1 \frac{d\theta}{d\lambda} \delta\lambda = \frac{f_1}{f_2} g \qquad (3.4)$$

We may then substitute in Equation (3.3) to obtain the flux entering the instrument and passing via the dispersing element into the image.

$$\Phi = \frac{BshA\tau_\lambda}{f_1^2}$$

Then the flux per unit area in the image is

$$\frac{BshA\tau_\lambda}{f_1^2} \div s \frac{f_2}{f_1} h \frac{f_2}{f_1}$$

$$= \frac{B\tau_\lambda A}{f_2^2} = \frac{B\tau_\lambda}{F^2}$$

where F is the focal ratio (f-number).

If we substitute for f_2 using Equation (3.4), this may then be written as

$$B\tau_\lambda A \left(\frac{d\theta}{d\lambda}\right)^2 \frac{\delta\lambda^2}{g^2} \qquad (3.5)$$

This analysis will enable us to compare the operation of a spectrograph with that of a monochromator, but first we might also note that for a continuum source the amount of flux in the bandwidth $\delta\lambda$ will be proportional to the bandwidth, so we may write $B = L\delta\lambda$ where L is the spectral radiance of the source. For a line source (by which in this context we mean that the natural linewidth is less than or equal to the resolution of the instrument), the incident flux is independent of bandwidth. We may, therefore, write for a monochromator or spectrometer:

$$\Phi = B \frac{d\theta}{d\lambda} A\tau_{\lambda\beta} \, d\lambda \qquad \text{(monochromatic)} \qquad (3.6a)$$

$$\Phi = L \frac{d\theta}{d\lambda} A\tau_{\lambda\beta} \, (d\lambda)^2 \qquad \text{(continuum)} \qquad (3.6b)$$

for monochromatic or continuum sources respectively. For a spectrograph limited by the spatial resolution of the detector, the irradiance is

$$I = BA\left(\frac{\mathrm{d}\theta}{\mathrm{d}\lambda}\right)^2 \tau_\lambda \left(\frac{\mathrm{d}\lambda}{g}\right)^2 \qquad (3.7a)$$

or

$$I = L\left(\frac{\mathrm{d}\theta}{\mathrm{d}\lambda}\right)^2 A\tau_\lambda \left(\frac{\mathrm{d}\lambda}{g}\right)^3 \qquad (3.7b)$$

When expressions (3.6) and (3.7) are normalized to the case of a source of unit radiance they describe the "luminosity" of the instruments.

The analysis as it stands is incomplete. It assumes that the slit is uniformly illuminated and would, therefore, be different for a point source. It does not take into account the practical problems of designing the optics required to focus a highly dispersed spectrum, nor does it take into account the factors such as the quantum efficiency and noise of the detector and the fact that the speed of photographic materials decreases with grain size. In choosing or designing the best instrument for his particular task the spectroscopist will have to take into account more factors than are included in expressions (3.6) and (3.7). However, this analysis is adequate for our present purposes because it enables us to separate the contributions from the various components of the system and to put into perspective the way in which the parameters of the grating influence the efficacy of the instrument.

In the expressions (3.6) and (3.7) we note that:

B and L are functions of the source;
$\mathrm{d}\theta/\mathrm{d}\lambda$, A and τ_λ are functions of the grating;
β (and F) are functions of the focusing and collimating optics although in Chapter 7 we shall see this can be incorporated into the grating;
$\delta\lambda$ is a function of the problem;
g is a function of the detector.

From the point of view of the grating it is the angular dispersion $\mathrm{d}\theta/\mathrm{d}\lambda$, the aperture, and the efficiency which influence the ease with which a spectrum is detected. The development of grating technology has been largely concerned with finding practical ways of maximizing these factors without introducing imperfections which would themselves limit the performance of this instrument.

Let us now take a closer look at the functions

$$A\frac{\mathrm{d}\theta}{\mathrm{d}\lambda} \quad \text{and} \quad A\left(\frac{\mathrm{d}\theta}{\mathrm{d}\lambda}\right)^2.$$

We saw from Chapter 2 that the angular dispersion $d\beta/d\lambda$ is found by differentiating the grating equation to be $m(d \cos \beta)$ and in the Littrow case we may write

$$\frac{d\theta}{d\lambda} = \frac{2 \tan \theta}{\lambda}$$

(which should not be confused with the expression relating to the rotation of the grating). If the area of the grating is A', then the aperture A is $A' \cos \theta$, as shown in Figure 3.3.

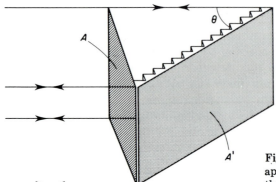

Figure 3.3   The effective aperture of a grating used in the Littrow mount.

So the product of aperture and angular dispersion now becomes

$$A \frac{d\theta}{d\lambda} = \frac{2 \tan \theta}{\lambda} A' \cos \theta = \frac{2A' \sin \theta}{\lambda} \qquad (3.8)$$

Since we are considering the Littrow case we may write

$$\frac{2 \sin \theta}{\lambda} = \frac{m}{d}$$

and

$$A \frac{d\theta}{d\lambda} = \frac{mA'}{d} = mNH \qquad (3.9)$$

where N is the total number of grooves, H is the height of the grating and NH represents the total length of the grooves. For a spectrograph the equivalent function is this multiplied by the angular dispersion, i.e.

$$A \left(\frac{d\theta}{d\lambda}\right)^2 = \frac{4A'}{\lambda^2} \sin \theta \tan \theta = \frac{mNH \tan \theta}{\lambda}$$

It follows from Equations (3.6) and (3.7) that the flux passing through a monochromator and the image intensity in a spectrograph are both proportional to the total length of the grating grooves. It is here that the interference grating offers a significant advantage over a ruled grating. There are two practical advantages. The first is that interference techniques are not limited by diamond wear and therefore the total length of groove is not limited, and the second is that (within certain limits) it is no more difficult to make gratings of fine pitch than coarse pitch. The latter does not, according to Equations (3.9) or (3.10), affect the total flux, but it does mean that a grating with a large total length of groove can be of a convenient size. Since interference gratings also have a much lower level of stray light than do ruled gratings, there is a very significant overall advantage in using them in monochromators and spectrometers.

It is interesting to note that in neither (3.9) nor (3.10) does the pitch of the grating appear. It is from this that the echelle derives its great advantage. It is designed to work at high angles of incidence and it is evident that for a given wavelength and a given area of grating, the flux though a monochromator and the image intensity in a spectrograph increase with this angle. In Figure 3.4 we plot the functions $\sin \theta$ and $\sin \theta \tan \theta$ as functions of θ and see that, in the case of a spectrograph limited by the spatial resolution of the detector, the increase of intensity is most dramatic for angles in excess of about $60°$. It is for this reason that those using spectrographs, particularly astronomers, have tended to use gratings in high diffracted orders. If we consider a $600\,\mathrm{groove\,mm^{-1}}$ grating of a given width used at $400\,\mathrm{nm}$, then one gains a factor relative to the first order and of 4.0, 9.5 and 18, by going respectively to the second, third and fourth order in a spectrograph. On the other hand, in a monochromator the gains are only 2, 3 and 4 respectively.

On might infer from these expressions that it is best always to use gratings at large angles of incidence and diffraction. This is true over a wide range of angles, but it is based on the assumption that as the dispersion increases the extra light allowed into the instrument through the wider entrance slit more than compensates for the fact that the angular subtense of the grating aperture is becoming smaller. In the limit the entrance slit width tends to infinity while the aperture tends to zero. This is obviously not a realistic configuration, first because a very wide slit will subtend a significant angle at the grating so that the angle of incidence is no longer unique and the theory breaks down. It would also become increasingly difficult to fill the slit uniformly. These considerations and the practical

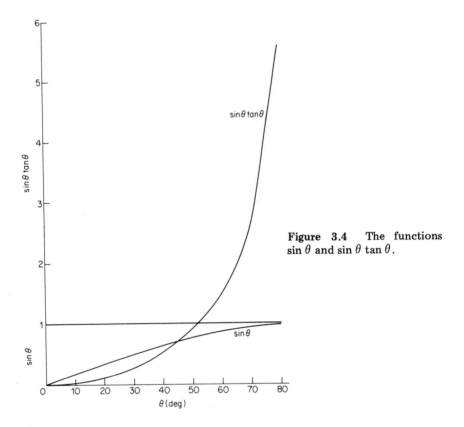

Figure 3.4   The functions $\sin \theta$ and $\sin \theta \tan \theta$.

difficulties of making a grating that will work efficiently at high angles of incidence and diffraction usually conspire to set the optimum blaze angle of an echelle at between 50° and 70°.

The whole of the analysis so far has been based on the assumption that either the total flux or the intensity of a spectral image should be maximized. Other factors must be taken into account. Ultimately the ability to detect a signal depends not on the level of the signal but on the signal-to-noise ratio. If one is working at very low light levels with an ideal detector, then the signal-to-noise ratio is determined by photon statistics and is equal to the square-root of the number of photons detected. However, in most cases the limit is set by other sources of noise such as detector noise or stray light from the grating and from other parts of the instrument. The intensity of ghosts and grass increases as the square of the sine of the angle of incidence (or as the square of the order number) and so too does the departure from flatness of the wavefront due to errors in groove position. Certainly, in the case of a ruled grating such as an echelle, the design of the optimum grating will take into account the tolerances that can be achieved on the ruling engines that are available. Furthermore, this result is only true for the Littrow angle. For wavelengths at different angles of diffraction the dispersion does not tend to infinity.

The other factor which must be taken into account is the free spectral range. Because of the limitation on the total length of groove that can be ruled with a diamond we have seen that the greater flux and intensity are obtained by ruling a comparatively coarse grating to be used at large angles of incidence in high diffracted orders. This imposes a severe limit on the free spectral range of the instrument and it is often necessary to employ "cross-dispersion" in which a further dispersing element is used to sort out the various overlapping orders. This will of course entail further losses but these are often more than compensated for by the increased dispersion, and hence intensity, that result from using the grating in higher orders. In some cases this is a positive advantage. The blaze condition is satisfied for different parts of the spectrum in different orders and with cross-dispersion a large range of spectrum can be displayed quite compactly, as shown in Figure 2.19. This is particularly useful in an astronomical spectrograph if the image is to be intensified electronically. Owing to the (apparent) weakness of the source one wishes to make best use of the light that is available, which implies the use of high dispersion, but at the same time the finite size of an image intensifier tube demands that the whole spectrum be focused into a comparatively small area.

In a monochromator or spectrometer in which the slits remain fixed and the grating is rotated, the use of an echelle with cross-dispersion for sorting the orders would be extremely inconvenient. Nor would it be justified, because we see from Figure 3.4 that the gain in transmission of the instrument is much less than the gain in irradiance in the spectrograph, so it is preferable to have as wide a free spectral range as possible and as we saw in Chapter 2 this implies the use of first order. Furthermore, in a monochromator one is not embarrassed by the fact that the first-order spectrum is dispersed over a wide range of angles. Since only one point of the spectrum has to be focused at any time, the focusing optics do not have to work over a large field and may therefore be simpler than in a spectrograph. However, in this type of instrument we must bear in mind that the angle of incidence varies with wavelength (whereas in an echelle spectrograph, by going to different orders it was kept within a small range of angles) and this will affect the transmission.

These arguments are quite appropriate when one has to design an instrument to work at a given wavelength and Equations (3.6) and (3.7) are very useful in helping to select or design the best possible grating in the best possible instrument. In practice, however, the design of an instrument may be influenced by many other factors, not least of which is the fact that it will have to work over a range of wavelengths. A grating optimized, for example, to give high dispersion and hence maximum flux in the blue may well have too fine a pitch to give any spectrum at all in the red. An instrument which is designed to work over a range of wavelengths cannot operate at the optimum performance over the whole of that range. Very large numbers of instruments, particularly monochromators for spectrophotometry, are built with small gratings and have a performance which falls far short of the best that our analysis would suggest is possible. This is because they are designed to give a required performance at the minimum cost and in fact the great majority of commercial instruments are built on this basis. They are a compromise between performance, usually over as wide a range of wavelengths as possible, the cost and availability of components, and other practical considerations such as the need to put the instrument into a box that can be transported without undue difficulty and which will pass through the door of the average laboratory. From Equation (3.9) we see that for a given diffracted order the grating required for the most efficient spectrograph would have the maximum total groove length. With an interference grating this is limited in principle by the coherence length of the laser but in practice by the size of the mirrors and blank and the available power of the laser. Let us be optimistic and assume that no expense is spared and that we have a grating with a diagonal (or diameter) of 500 mm. If we assume that the focusing and collimating optics have a focal ratio of $f/8$ (again an optimistic assumption) then the focal length will be 4 m. It is therefore unlikely that the instrument will be much less than 5 m long particularly when we consider that the grating will probably weigh 200 kg and will therefore require very substantial metal mounts and bearings which will weigh far more than this. For example, the grating test bench recently built at the National Physics Laboratory will accomodate gratings up to 300×250 mm. It has a focal length of 6 m with a folded optical path, so the total length of the instrument is 4 m and it weighs approximately $4\frac{1}{2}$ tonnes and was assembled *in situ*. Clearly such an instrument would be impractical from a commercial point of view and rather than go to these lengths most spectroscopists would find it more profitable to devote their time and money towards improving the detector or making the source brighter. In astronomy, of course, one can do very little about the source except build a bigger telescope to collect more light from it. This in itself if very expensive and in the case of the weaker stars the cost of a spectrum may be over £1 per recorded photon. Under these circumstances one does have to use the biggest available gratings and at some observatories the spectrographs are vast.

☐ Comparison of a grating and a prism

Our analysis as far as Equation (3.4) refers equally well to any type of dispersing element and in particular it also refers to prisms, so we may use it to compare the performance of prisms and gratings. If we consider the case of a monochromator then the factors to be compared are

$$B\left(\frac{d\theta}{d\lambda}\right)A\tau_\lambda\beta\delta\lambda_{\text{prism}} \quad \text{and} \quad B\left(\frac{d\theta}{d\lambda}\right)A\tau_\lambda\beta\delta\lambda_{\text{grating}}$$

In the case of the prism it may be shown (Jaquinot 1954) that the expression $A(d\theta/d\lambda)$ is equal to $A_P(d\mu/d\lambda)$ where A_P is the area of the base of the prism.

In the case of the grating we see from Equation (2.11) that

$$\frac{d\beta}{d\lambda} = \frac{m}{d\cos\beta}$$

In the Littrow mounting (which we consider for simplicity, in order to avoid anamorphism) we may write

$$\frac{d\theta}{d\lambda} = \frac{2\tan\theta}{\lambda}$$

from which it follows that

$$A\frac{d\theta}{d\lambda} = \frac{2A_g\sin\theta}{\lambda}$$

where A_g is the area of the grating.

If, in a typical case we set $2\sin\theta = 1$ the ratio of the lumin-

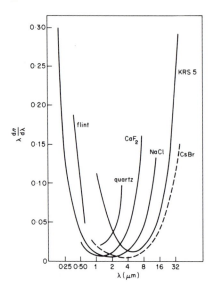

Figure 3.5 The variation as a function of wavelength of the quantity $\lambda(d\mu/d\lambda)$ for various materials (after Jaquinot 1954).

osities of prism and grating instruments of the same effective resolving power is given by the expression:

$$\frac{(d\theta/d\lambda)A\tau_{\lambda(\mathrm{p})}}{(d\theta/d\lambda)A\tau_{\lambda(\mathrm{g})}} = \frac{(d\mu/d\lambda)A_\mathrm{P}\tau_{\lambda(\mathrm{p})}}{(A_\mathrm{g}/\lambda)\tau_{\lambda(\mathrm{g})}} \qquad (3.11)$$

If we further assume the area and efficiency of the grating to be the same as that of the prism, the expression then reduces to $\lambda(d\mu/d\lambda)$, which is a function of the material of the prism and for which there is no simple analytical expression. Measured values of this expression for a variety of common prism materials are shown in Figure 3.6 and from this we can see that prisms are generally between 1% and 30% as effective as gratings.

Bibliography

For an excellent introduction to the design of spectroscopic instruments see:

Bousquet, P (1971). "Spectroscopy and its Instrumentation" (translated by Greenland, K. M.). Hilger, London.

For a more detailed consideration of the luminosity of spectroscopic instruments see:

Jaquinot, P. (1954). *J. Opt. Soc. Am.* **44,** 761
and
Stroke, G. W. (1967). pp. 486—504.[†]

[†] See main references.

4

The Manufacture of Gratings

The classical method of manufacturing gratings, the method used for over 150 years, is to scribe, burnish or emboss a series of grooves upon a good optical surface. Originally this surface was one of highly polished speculum metal but ever since vacuum-coating techniques were developed by Strong in the 1930s the majority of gratings have been ruled in thin layers of aluminium or gold deposited upon a glass substrate.

There are basically two problems in ruling gratings, the first is to produce consistently a groove of the right shape and the second is to ensure that it is put in the right place. As we saw in Chapter 2, the shape of the groove determines the efficiency of the grating and its quality (degree of surface roughness and jagged edges) will influence the level of diffuse scattering. If there are variations in depth or shape along the length of the groove, these can generate stray light. The density of the grooves determines the dispersion and the accuracy of position of the grooves determines the quality of the spectral image. Long-term errors in position affect the flatness of the diffracted wavefront and hence distort the spectral image. This can both reduce the resolving power of the grating and increase the level of light diffracted in between the main orders. Short-term errors also contribute to the level of light that is to be found between the orders. Periodic errors, as we have seen, generate ghosts, and random errors generate a general background scattering known as grass. The problem of creating the right shape of groove is very important, but in principle it is generally less crucial than that of putting the grooves in the right place. A grating with an inferior

71

efficiency can in most cases still be used, but a grating with a distorted wavefront will never give the full resolving power; and a high level of stray light will set an upper limit to the signal-to-noise ratio that a given instrument can achieve.

□   The formation of grooves

A parameter which is sometimes quoted and which helps to put into perspective the magnitude of the task of ruling a grating is the total groove length. A typical grating might be 100 mm wide with grooves 100 mm long and have 1200 grooves mm^{-1}, in which case the total groove length is 12 km. Such a grating is not exceptional, but it would be among the largest currently used in commercial instruments. For more specialized instruments, such as those used in astronomy where high light gathering power and/or high resolution are required, much longer or much finer gratings are called for. The groove is formed by the burnishing action of a diamond tool without the removal of material. Pressure on the tool is often very large. (It is difficult to specify or measure but may be of the order of tonnes per square centimetre.) Furthermore, this pressure is not constant, since it is applied and removed as each individual groove is formed, so the stresses on the diamond are very severe. Natural diamond has an octahedral structure with an angle of 79° between the (111) and ($\bar{1}\bar{1}\bar{1}$) planes, but the tool is usually polished to an angle of 90° or more in order to give the required groove shape. In order to attain maximum hardness it is necessary carefully to align the tool with respect to the crystallographic axes. Even so, the life of any given diamond cannot be predicted. The frustration of finding, after say ten days of ruling, that the diamond fractured half-way through is one of the occupational hazards of ruling gratings.

There are two common forms of ruling diamond. The first is a straight chisel or roof shape and the groove is formed by dragging the edge of the tool through the metal. This type is comparatively easy to align and since only a small area is in contact with the metal it requires a comparatively light loading. However, the stresses on it are concentrated upon a smaller area and when it does fracture or wear, it must be discarded. The second type is boat-shaped and the ridge of the tool is a curve defined by the intersection of two cones. This has the advantage that different sections of the tool may be used (on different gratings!) and its useful life extended, but it does require far greater loads and its alignment is more critical than

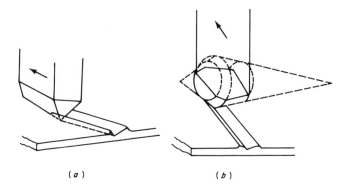

Figure 4.1 The two basic types of ruling tool: (*a*) double-ended (roof edge tool); (*b*) double-cone tool consisting of intersecting conical surfaces.

with the roof type. The two forms of tool are shown schematically in Figure 4.1.

When setting up to rule a grating, great care is required to ensure the correct alignment of the tool and its correct loading. The shape of the final groove is not simply the same as that of the tool but is the result of a complex process in which each groove interacts with its neighbour. Since no metal is removed during the ruling process, the displaced material forms a ridge on either side of the groove and the correct alignment of the edge of the tool with respect to the groove (the twist setting) is achieved when these two ridges are of similar height. Once this alignment is established, it is necessary to determine the correct loading of the tool. If the load is too light then there will be unruled areas or "land" between the grooves and if it is too great the groove is distorted, diamond wear is increased and so is the possibility of the tool sticking to the material. When a series of grooves is ruled, the shape of a given groove is changed as the subsequent groove is ruled. Material will be pushed away from the tool and this will tend to steepen the angle of the adjacent facet. It is therefore preferable to arrange that each new groove is ruled on the side of the steep facet of the existing grooves. The steep facet will change and will reduce the included angle of the groove, but it will not affect the angle of the shallow facet which is usually the most important from the point of view of the blaze. The correct loading of the tool is generally taken to be that which yields the best approximation to a saw-tooth groove profile. The measurements of these profiles and the setting up of the diamond is an art which is based largely on experience; particularly so with gratings of fine pitch. The resolution attainable from an optical microscope is often inadequate to give a clear picture of the groove and electron microscopy is very tedious. However, useful information can be obtained very rapidly with the use of a Talystep which traces out the profile of a surface by drawing a lightly loaded diamond stylus across it (Verrill 1973, 1975). A series of Talystep traces are shown in Figure 4.2 and illustrate the various features of the operation of setting up the ruling tool.

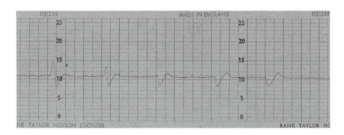

Figure 4.2 Traces of singly ruled grooves; trials for the orientation setting of the ruling tool; five twist settings at intervals of 15 minutes of arc. Vertical scale 500 nm, horizontal scale 5 μm.

At the extremes of the ruling process further difficulties are encountered. For very shallow grooves the shape of the grooves is greatly influenced by the elastic recovery of the material. In all cases the shape of the top of the groove is the product of ruling two facets, whereas the bottom of the groove is generated by the tip of the ruling diamond and has a shape which is more closely controlled. Many gratings are used away from the Littrow mounting so that the bottom of the groove is in shadow. Under these circumstances it is often found that a first- or other odd-generation replica has a higher efficiency than the original master. This is because the shape of the apex of the groove is now that defined by the ruling diamond. This is particularly important in the case of echelles, where either one has to take great care to adjust the tool and the loading to give a sharp apex or one has to restrict oneself to the use of odd-generation replicas.

Practically all modern ruled gratings are ruled in a layer of metal, usually aluminium or gold, which has been vacuum-deposited upon the surface of an optically flat glass substrate. The most notable exception to this rule are gratings destined for use with high-powered infrared lasers in which case the grooves are ruled directly onto a substrate of, say, Kanigen copper that has been designed for the rapid extraction of heat. Aluminium has the advantage that it is readily available and is comparatively easy to deposit in a uniform, homogeneous and strongly adherent film. It has a high reflectance over a wide range of wavelengths, which is often convenient for testing purposes, but is of no direct advantage unless the master itself is to be used in an instrument. The disadvantage of aluminium is that, although the metal itself is fairly soft, it always forms a layer of oxide on the surface. This layer is typically about 3 nm thick, is far harder than the alumium and is responsible for a high proportion of the wear suffered by the ruling diamond.

The favourite alternative to aluminium is gold. It has the advantage that it is soft and chemically inert so that it does not tarnish. It is, however, particularly difficult to deposit a film of gold which will adhere strongly to the substrate. In practice it is necessary first to deposit a film of some material which does adhere strongly to the glass and has an epitaxial strcture compatible with gold. Chromium films are often used for this purpose, but it still requires some skill to produce a gold film that withstands the enormous stresses of ruling. The other drawback with gold is its cost. The amoung of gold actually deposited on the grating blank is quite small and even for a large echelle requiring a thick film is only of the order of a few grammes. However, the process of depositing a film of uniform thickness is extremely wasteful because the greatest part of the material is deposited on the inside of the coating plant.

For gratings which are used in first order the thickness of the film should be about the same as the blaze wavelength or slightly less. If the film is to be flat to 1/20 of the wavelength then this implies that the thickness must be uniform to within about 5%. (For an echelle, which might require a film 25 μm thick, the thickness must vary by less than 1%.) The uniformity of film thickness may be increased by increasing the distance between the evaporation source and the substrate. In fact the thickness varies with the angle of incidence according to a $\cos^3$ or $\cos^4$ law depending upon the source. If we assume a $\cos^3$ law then for a 5% uniformity, the substrate can subtend an angle of $\pm 10°$ and for 1% uniformity only $\pm 4°$. If the evaporation source emits into a solid angle of 1 steradian this implies a coating efficiency of 3.3% and 0.7% respectively. It also implies that for a blank with a diagonal of 300 mm the coating plant must be about 0.8 m long and 2 m long respectively. In practice the working distances are significantly reduced by placing the source to one side and rotating the blank or by evaporating the metal from a series of sources. However, despite these techniques, the process is still wasteful of material.

The ruling process demands far more from vacuum-coating techniques than is usually encountered with other applications. The thickness of film is often far more than would be required for, say, a mirror. The quality of the grating depends not so much upon the optical quality of the the film but upon its mechanical properties (indeed the optical quality, particularly the surface smoothness, is often improved by ruling) and the tolerances on imperfections are far more severe. In the case of a mirror the effect of a small speck or globule of metal is restricted to the area it occupies. On a grating it will affect one or maybe several grooves and may even be picked up by the diamond and interfere with the ruling over a significant area of the grating. It could even cause the diamond to fracture. Hence, despite great advances in the field of vacuum-coating technology, the difficulties of coating a blank for ruling should not be underestimated.

□   Mechanical tolerances

The problem which has received by far the most attention is that of putting the grooves in the correct place. The tolerances involved are probably the most severe of any mechanical operation and in some cases demand accuracies of the order of atomic dimensions. If the grating is to have the resolving power predicted by the theory, then the diffracted wavefront must be flat. The Rayleigh tolerance suggests that it should not deviate from flatness by more than $\lambda/4$, although $\lambda/10$ is the tolerance that is usually aimed for. This means in effect that all parts of the grating should diffract the incident wavefront into precisely the same direction and this in turn means that the pitch should be constant over the whole of the grating.

Let us for the sake of simplicity consider a grating to be used in the Littrow mounting at a wavelength λ. If a groove is out of place by an amount δx then the contribution from this groove will have travelled an extra distance of $2\delta x \sin \theta$. If the wavefront is to be flat to $\lambda/10$ then it follows that

$$\delta x < \frac{\lambda}{20 \sin \theta}$$

Since in this mounting $2d \sin \theta = m\lambda$ we may also write

$$\delta x \; < \; \frac{d}{10m}$$

so that on this basis a given grating would be expected to have the same resolving power in a given order irrespective of wavelength. In practice the performance depends rather more on the form of the wavefront error. However, for the purpose of establishing the tolerance as groove position this analysis is sufficient to tell us that for a 1200 groove mm^{-1} grating used in the first order the maximum error should be about 80 nm, whereas for an echelle used in the visible and blazed at 65° the tolerance is about 25 nm. So although the pitch on an echelle is much coarser, the tolerance on groove position is in fact tighter than for a conventional first-order grating. These tolerances refer of course to the final error in the position of the groove. Such an error may arise from a variety of sources such as wear, distortion or thermal expansion, and it follows that the error due to each of these must be substantially smaller than this if the final error is to remain within tolerance. Not only must the grooves be in the correct position but they must also be parallel. A small but constant angle between consecutive grooves is known as fan error and in effect means that the pitch of the grating varies from one end of the groove to the other. This in turn means that the spectral image for such a grating is the sum of a series of images each slightly displaced, corresponding to different positions along the groove. In the case of the echelle quoted above, if the grooves are 150 mm long it follows that they must be parallel to within

$$\frac{25\,\mathrm{nm}}{150\,\mathrm{mm}} \; = \; 1.7 \times 10^{-7}\,\mathrm{radians} \quad (\text{or } 0.03 \text{ seconds of arc})$$

The tolerance on the straightness of the grooves is less severe provided that all the grooves have the same shape. In this case the spectral images corresponding to different positions along the groove will be superimposed (since each slice of grating has the same pitch). The resultant image will be degraded along the direction of the slit, but this will have less effect on the resolution. An acceptable departure from straightness of the grooves would be about

$$\frac{3d}{10m}$$

The fact that the grooves need not be absolutely straight is particularly important because they may then be ruled with a diamond which moves with simple harmonic motion without the need to stop the sideways movement of the grating blank. This results in considerable simplification in the design of some ruling engines.

As we have seen, any periodic errors in the positions of the grooves will generate ghosts or false spectral lines. The intensity of such ghosts relative to the main diffracted order is given approximately (Rowland 1893) by

$$\frac{I_{\text{ghost}}}{I_{\text{order}}} = \left(\frac{\pi\,\delta x\,\sin\theta}{\lambda}\right)^2 = \left(\frac{\pi\,m\,\delta x}{d}\right)^2$$

A value of 10^{-6} or less would be a reasonable figure to aim at from the spectroscopic point of view so that

$$\delta x < \frac{\lambda}{10^3\,\pi\,\sin\theta} = \frac{2d}{10^3\,\pi m}$$

in the Littrow case.

For a grating of 1200 grooves mm^{-1} used in first order this gives a value of about 0.5 nm, (5 Å).

If there are random errors in the position of the grooves, then these will generate a series of random ghosts which contribute a background level of light called "grass" scattered in between the spectral orders. It is called grass because, on a photometric trace of the light intensity, that is just what it looks like. The proportion of incident light thus scattered (Maréchal 1958) is given by

$$\left(\frac{4\pi\,\overline{dx}\,\sin\theta}{\lambda}\right)^2$$

where $\overline{dx}$ is the RMS error in groove position. It is important to reduce this figure as far as possible, since for practically all ruled gratings it is this which sets the upper limit to the signal-to-noise ratio that an instrument can achieve. However, if we accept a figure of 1% as being realistic it follows that the RMS error should not exceed

$$\frac{\lambda}{120\sin\theta} \quad \text{or} \quad \frac{d}{60m}$$

which, for a 1200 grooves mm^{-1} grating used in first order is about 1.3 nm.

The task of ruling a grating can be stated in terms of ruling a single groove equivalent to the total groove length. The groove may be 20 km or more long, it must have an overall straightness of perhaps 80 nm, random short-term deviations must be less than 1.5 nm and periodic deviations from straightness must be less than 0.5 nm. It is not, therefore, surprising that it was over 100 years after Fraunhofer's work that gratings became available to more than just a handful of spectroscopists.

☐ Ruling engines

The basic design of all ruling engines is similar in that all are required to produce two orthogonal motions at vastly different speeds. The fast motion as the diamond moves along the length of the groove may be at a speed of perhaps 100 mm per second, and the far slower motion as the diamond progresses from one groove to the next may typically be at a speed of perhaps 100 mm per week. There are several different designs of machine which will achieve this. The diamond may, for example, reciprocate along a fixed carriage while the blank is slowly moved across. Alternatively, the blank may remain stationary while the whole diamond carriage moves across or, again, the diamond may advance slowly from one groove to the next while the blank reciprocates beneath it. These are shown diagrammatically in Figure 4.3. The first, the so-called "shaper" design was that developed by Rowland (1882). The second, the "planer" design, was developed by Strong (1951) and the third was proposed by Stroke (1967). The choice of design will depend largely upon the type and size of gratings required. In general it is preferable to move only the lightest components at high speed and the heavier components more slowly, in order to reduce the inertia of the system. This reduces the out-of-balance forces that tend to distort the engine and reduces mechanical vibration. For very large gratings the blanks themselves become very heavy and some of the largest that have been proposed would weigh 200 kg or more. For these gratings the diamond carriage may weigh less than the blank, and it is better to keep the blank stationary. For smaller gratings the reverse is true and the blank is moved in at least one direction. The most common form of ruling engine is that in which the diamond reciprocates along a fixed carriage while the blank is translated slowly underneath. The advantage of this type is that it satisfies the conditions for minimum inertia. However, engines have been built in which the blank oscillates and the diamond carriage moves slowly. This type has the advantage that fanning of the grooves is automatically eliminated.

Most ruling engines are of the type shown in Figure 4.3(a) and most of what follows will be expressed in terms of this design,

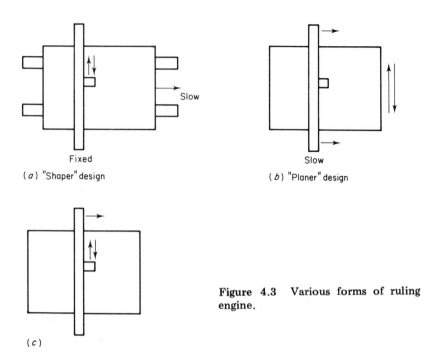

(a) "Shaper" design

(b) "Planer" design

(c)

Figure 4.3 Various forms of ruling engine.

although from the point of view of the grating it is only the relative motion of the blank and the diamond that matters.

The next choice which faces the designer of a ruling engine is whether the motion of the blank should be continuous or whether the blank should stop while the groove is being ruled and then move on to the next position. The advantage of the stop–start design is that since the blank is nominally at rest during ruling, the straightness of the groove does not depend upon the form of the reciprocating motion of the diamond. On the other hand, if the blank moves continuously and the diamond reciprocates under simple harmonic motion, then the grooves will be S-shaped rather than straight. This is often acceptable since the tolerance on straightness is not as severe as the tolerance on groove position. In gratings intended for first-order use, the acceptable deviation from straightness may be as much as 0.3 of the grating spacing and this means that 80% of the stroke of the diamond can be used for ruling. However, for gratings to be used in higher orders, the useful fraction of the stroke becomes much smaller, the total excursion of the diamond must then be much greater than the length of the groove and for large gratings this implies a large engine with long pushrods which would both be cumbersome and prone to vibration. The problem is most extreme in the case of large-aperture coarse echelles where it is essential to linearize the motion of the diamond (Harrison and Stroke 1955). This is often regarded as a small price to pay for the advantages of moving the blank continuously. If the blank carriage does not have to stop, then variations in friction and in elastic deformations are greatly reduced and the forces acting on the engine are more or less constant. The inertia of the blank becomes an advantage since it helps to smooth out any variations in friction, but in a stop–start machine it is an embarrassment. It is also very much more difficult to control the blank carriage when it has stopped than to control its uniform motion.

Most of the early engines, those due to Rowland, Blythswood and Michelson for example, were of the stop–start variety, but most modern engines employ continuous blank motion either with or without some means of linearizing the motion of the diamond. Some of the most demanding gratings from the point of view of the straightness of groove (and from other aspects) have been large echelles ruled on the interferometrically controlled engines built at the Massachusetts Institute of Technology. These were ruled with continuous blank motion. There are, on the other hand, some occasions when a stop–start machine may be the most appropriate as, for example, when Bausch & Lomb (Loewen 1972) ruled a 125 mm square grating with 3600 grooves mm^{-1} and a facet angle of only 3°. This demanded a very light loading on the diamond but maximum freedom from vibration and it was found that one of Michelson's engines was the best from this point of view as it gave greater stability.

The most common means of generating the slow motion, either of the grating blank in the Rowland type of engine or of the diamond carriage in the Strong design, is by a precision screw. In all engines built prior to 1955 this screw served the dual purpose of providing the slow drive and of defining the position of the groove. The screw was attached either directly or via gears to a dividing head which had to be made with great precision, and it was this that determined the pitch of the grating. Thus, for example, if the lead screw had a pitch of 40 threads per inch and was connected directly to a dividing head of 360 teeth then the carriage would advance by 1/14 400 inch as the screw was rotated by one increment of the dividing head. This would then generate a grating of 14 400 grooves per inch (or 578 grooves mm^{-1}), which was a very common pitch for early gratings. The pitch could be varied by connecting the dividing head to the screw via a series of gears, but the introduction of gears brought with it the possibility of more periodic errors. In order to attain sufficient accuracy of groove position it was necessary to take the utmost care in producing the lead screw. This was generally first cut on a lathe and then lapped with a split nut which was constrained to move along straight ways. The action of lapping removed material from both the nut and the screw in such a way that irregularities were reduced. The nut was made adjustable so that it could take up the slack due to the removal of material and so continue the lapping process until a very precise screw had been generated. However, as Rowland pointed out it is more difficult to mount a screw than it is to make it. The bearings must allow no longitudinal shift because this would result in an error of the groove position. It is therefore necessary to maintain a constant thickness of oil films between the bearing surfaces for the duration of the ruling operation. Similarly, any variation in the thickness of the oil film between the nut and the lead screw must be kept to a minimum because this will introduce variations in the position of the grooves. The bearing surfaces on the screw will not generally be concentric with the helix. They may well have been concentric when the screw was first cut, but the process of lapping causes the helix to become eccentric and this eccentricity can lead to a periodic error in the position of the grooves and hence to a ghost in the spectrum. Care must be taken to reduce the sag of the lead screw because this too will lead to unacceptable errors in groove position. The sag can be reduced by a factor of 50 by the addition of suitable counterweights, but to do this requires a thrust bearing which permits the shaft to extend beyond the bearing in order to accomodate the counterweights: Strong (1951), for example, achieved this by the use of a "zero lead" screw as shown in Figure 4.4.

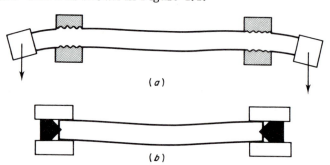

Figure 4.4 The effect of the bearing and counterweights on the sag of a lead screw: (a) zero lead screw, (b) thrust bearing.

However much care is taken with the manufacture and the mounting of the lead screw(s), there will always be some residual errors. The limit is set as much by the materials that are available as it is by the skill of the engineer or craftsman. There will always be some measure of creep and some amount of wear in use so that even a perfect component could not be guaranteed to stay that way. Much of the skill of building a ruling engine lies in the art of designing a system which is as far as possible immune to the inevitable imperfections of material and workmanship, and then employing both materials and workmanship that are of the highest standard. With these imperfections in mind Michelson (1927) took particular care to measure the errors in his lead screw and to design a cam which would compensate for them. The operation could then be repeated as often as was necessary in order to compensate for wear and creep. This particular engine was subsequently taken over by Bausch & Lomb and was in operation 50 years after it was built. By this time it had become so stable that it was only necessary to rework the cam once in four years.

The screw is, on the face of it, a very convenient device because it provides both the thrust necessary to drive the blank carriage (or the diamond carriage in the planer design) and a means of indexing the amount of movement from one groove to the next. However, we have seen that the mechanical tolerances imposed upon the screw are so severe that it is impractical consistently to rule high quality optical gratings on an engine of this type. It was therefore a significant advance when Harrison and his team at MIT introduced interferometric control to a ruling engine in the 1950s (Harrison and Stroke 1955). In effect this separated the two functions of driving the blank and controlling its position and greatly reduced the dependence of the quality of the grating upon the quality of the screw. This did not mean that it was possible to use a lead screw of inferior quality, rather it meant that it was possible to make better gratings using machines made to the highest standard of workmanship.

There are, as we shall see, practical difficulties introduced by the use of interferometric control, but it is a sound principle of precision engineering to separate the two functions of drive and guidance. Not all of the engines which were fitted with interferometric control immediately produced superior gratings and there are some occasions where, on balance, it is better to use a purely mechanical engine. However, the best ruled gratings which have so far been produced would not have been possible without the use of interferometric control. It is probably also true to say that the majority of ruled gratings in use today were produced on engines of this type.

If the position of the grooves is defined interferometrically, the function of the screw is merely to translate the blank with respect to the diamond carriage. There are other drive mechanisms which might equally well fulfil this function and it is interesting to note that two of them have in fact successfully been incorporated into the design of ruling engines. The first is a hydraulic system developed by Horsfield (1965) and the second is a piezoelectric inching mechanism developed by Gee (1975) and by Bartlett and Wildy (1975).

In the hydraulic system oil is pumped by a small reciprocating pump into a cylinder which rides on a stationary ram attached to the bed of the machine. The cylinder is attached by a ball-ended rod to the blank carriage which rides along precision-lapped steel ways and which is partially supported by a bath of mercury. The position of the blank is measured by means of a Michelson interferometer. The signal from the interferometer is measured photoelectrically and is used to servo control the pump. The whole system is extremely simple, avoids all the periodic errors associated with leads screws and gears and has produced gratings with extremely low levels of stray light. It also has the advantage that the thrust is applied in the plane of the blank so that no bending stresses are generated as the carriage is advanced.

The drive for a piezoelectric system consists basically of a cylinder of piezoelectric material with electromagnetic clamps at either end. One end is clamped and a voltage applied to the cylinder which expands. The other end is then clamped and the first released, the voltage is turned off and the cylinder returns to its original size. The first end is then clamped and the process repeated and in this way the device inches along a cast iron vee-way. Again the position of the blank is monitored by a Michelson interferometer, the output of which is used to control the voltage applied to the piezoelectric element and thus to maintain the blank in the correct position as the groove is ruled. The majority of the weight of the blank carriage is borne by rafts floating in troughs of oil and the system is so adjusted that the required drive force is reduced to the equivalent of about 0.05 N.

☐　　　The control of the groove position

Basically there are two systems which have been used to measure and thereby to control the position of the blank relative to the ruling diamond to ensure that the grooves are ruled in the right place. The first is some form of optical interferometer of which the Michelson interferometer is the most common. The second involves the use of the moiré fringes that are produced when a transmission grating is laid over another grating of similar pitch.

The basic Michelson interferometer is shown in Figure 4.5(a). A beam of light from a suitable monochromatic source (originally a single-isotope mercury lamp but now almost always a single-frequency, stabilized helium–neon laser) passes through a beam

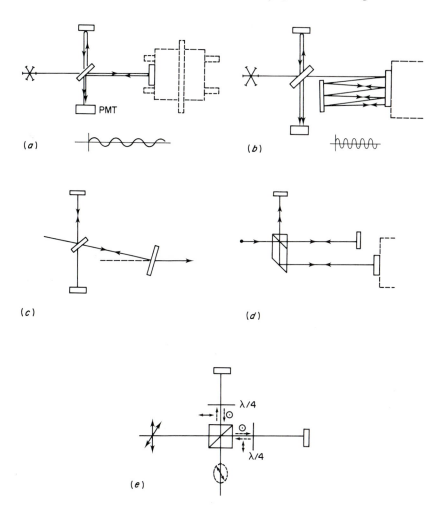

Figure 4.5 Various forms of interferometer for the control of a ruling engine.

splitter. One beam is reflected back down its own path by a plane stationary mirror, the other is similarly reflected back down its own path by a plane mirror attached to the grating blank. Parts of the two beams are then recombined at the beam splitter and are detected by a photomultiplier. If the optical path lengths of the two beams are equal, or differ by an integral number of wavelengths, then the contributions from the two beams are in phase and will interfere constructively to give a maximum intensity. If the path difference is an odd number of half wavelengths then destructive interference occurs and the intensity is zero. As the blank carriage is translated the output varies sinusoidally at a frequency which corresponds to one oscillation for a movement of half a wavelength. In its simplest form an engine could be made to count a given number of fringes and rule a groove when the interferometer output corresponded to a recognizable part of the sine wave. For example, if the engine ruled a groove on every fifth occasion that the intensity reached its minimum value, then if the light source was a helium–neon laser the pitch of the grating would be $1.57\,\mu m$ or $632\,grooves\,mm^{-1}$. Measuring the minimum intensity has the advantage that the signal is less dependent upon fluctuations of intensity of the light source, but is has the disadvantage that at this point the sensitivity of the system (i.e. change of signal level per distance moved) is a minimum and indeed is zero at the setting point. The system is much more sensitive if a position of maximum slope is chosen but since this occurs at half the maximum intensity it is necessary to make provision in the electronics for fluctuations in the brightness of the source. Harrison and Stroke (1955) used a system in which a series of gears driven synchronously with the diamond carriage generated a signal of a frequency which was nearly equal to the interferometer output. The two sinusoidal signals were then balanced for amplitude and kept in phase by an electronic system which acted to control the motion of the blank. In this case the frequency of the monitoring signal was 2.6 Hz. In the case of an engine ruling a $1200\,grooves\,mm^{-1}$ grating at 12 strokes per minute the grating blank travels at $0.17\,\mu m$ per second, which with red light would generate a frequency of approximately 1 Hz. It is very difficult to design electronics which will work at this frequency (that is, effectively DC) without drifting, particularly over the sort of periods required for ruling a grating. The frequency can be increased by incorporating multiple reflections into the beam as shown in Figure 4.5(b) but this system is more sensitive to misalignment and demands a light source with a coherence length which is several times the width of the grating. The system can be made insensitive to misalignment by the use of roof or cube-corner reflectors. Modern lasers have adequate coherence, but nevertheless multiple-pass systems are not common and it is more usual to modulate the beam at high frequency in order to simplify the electronics.

If, as is usually the case, it is required to rule at a pitch which does not correspond to an integral number of fringes, then it is usually necessary to incorporate a train of gears between the motion of the blank and the motion of the diamond carriage. Gears are always liable to give rise to periodic errors that generate spectral ghosts and an alternative system used by Horsfield was to incline the axis of the interferometer to the direction of motion of the blank carriage, as shown in Figure 4.5(c). In practice it is better if the reference arm of the interferometer is bent as in Figure 4.5(d) so that the two beams are parallel and as close together as possible. In this way it is possible to mount the reference flat on the diamond carriage itself and, since much of the paths of the beams are common, the sensitivity of the system to turbulence or temperature gradients is reduced.

An interesting variation of the Michelson interferometer is the polarizing version (Dyson *et al.* 1970) shown in Figure 4.5(e). In this case unpolarized light, or light polarized with its electric vector at $45°$ to the plane of the interferometer, is split with a polarizing beam splitter, so that the polarization of the transmitted component is perpendicular to the reflected component. Both beams then pass through quarter-wave retardation plates. Since each beam passes twice through its plate the net effect is to rotate its plane of polarization by $90°$. The transmitted beam is then reflected at the beam splitter and the reflected beam is transmitted on its return. The combined beam then consists of two components of equal amplitude but of different phase depending upon the difference in path lengths in the two beams. This then generates elliptically polarized light which, when passed through a further quarter-wave plate, is converted into linearly polarized light, the orientation of which varies as the blank carriage moves. It is then possible to servo control a rotating analyser to be crossed permanently to the output of the interferometer to within $1/6°$. This corresponds to a sensitivity of $1/1000$ fringe or for red light, a linear displacement of about $0.5 Å$ ($0.05 nm$) which is far better than could be achieved with a simple Michelson interferometer. It also has the advantage that it is more efficient since nominally 100% of the light is used rather than only 25% and it is insensitive to fluctuations of intensity of the light source.

Such an interferometer has been successfully applied to a ruling engine of 250 mm capacity at the National Physical Laboratory. Here the rotation of the flywheel driving the reciprocating diamond carriage is connected via a series of gears to the rotating analyser and its rotation is servo controlled to keep the analyser in step with the output of the interferometer. This in itself is an interesting deviation from the normal mode of control. In most engines the position of the blank is monitored and servo controlled by the interferometer. In this case the position of the blank is monitored but it is the motion of the diamond carriage which is servo controlled. This has the advantage that the control is applied to the lighter and faster moving component. The response of the system is therefore faster and is not troubled by slip stick and the other problems associated with control at very slow speeds.

An entirely different system of interferometric control was developed by Gerasimov and colleagues at the State Optical Institute in Leningrad (Gerasimov 1967). This makes use of the moiré fringes that are obtained when one grating is laid over another. The use of gratings to generate moiré fringes for the measurement of displacement is discussed in more detail in Chapter 9, and usually the fringes are formed by the superposition of two similar gratings which consist of alternate light and dark strips.

However, the system used by Gerasimov employs one transmission grating and one reflection grating of half the spatial frequency of the first. The arrangement is shown in Figure 4.6 and in this context it is perhaps appropriate to consider the similarity between this and the Michelson interferometer. The transmission grating fulfils the function of the beam splitter/recombiner and the reflection grating serves, in different orders, as the two mirrors. The relative phase of the two returning beams depends upon the relative lateral position of the two gratings, so that as one grating is moved with respect to the other the resultant intensity varies cyclicly with one cycle corresponding to a displacement of one groove of the transmission grating. However, unlike the Michelson interferometer the optical path difference is very small, so that it does not require a highly coherent light source. The pitch of the gratings is so chosen that the angular dispersion of one grating compensates for that of the other, so the system is independent of wavelength and will operate with a broad-band source. Furthermore, because of the small optical paths involved, the system is immune to changes in barometric pressure.

The use of moiré fringes in this way is in effect the optical

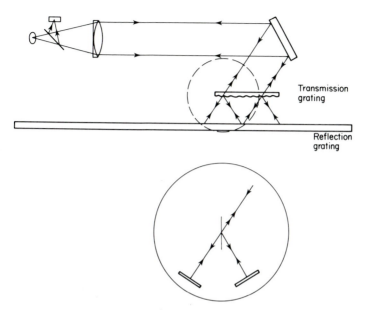

Figure 4.6 The use of moiré gratings to control a ruling engine.

equivalent of the Merton Nut because one starts with a pair of gratings which were ruled mechanically and uses them to control the ruling of the next generation of gratings. As with the pith nut in the Merton–NPL process, the effects of short-term errors are greatly reduced on the final grating because the position of any groove is determined by an average position of quite a large number of grooves on the control grating. As with engines controlled by a Michelson interferometer the intensities of ghosts from the gratings are dramatically lower than those from gratings ruled on a purely mechanical engine.

☐   The servo control system

The response time of the servo system should be chosen with some care. If it is too short, then there is a danger that oscillations will occur and this will give rise to diamond "chatter" and ultimately to increased levels of scattered light in the grating. If it is greater than one-tenth of the time taken to rule a groove, then this reduces the improvement in the overall accuracy of the groove position because much of the correction will be applied after the diamond has been raised. In most cases the diamond is actually ruling for about one-third of the ruling cycle time, so the servo response time needs to be no greater than one thirtieth of the period of reciprocation of the diamond. In an engine in which the blank carriage is moved continuously, if an error occurs which puts the blank ahead of where it ought to be, then the servo will respond by slowing down, halting or even reversing the motion of the blank. If the response time is one-thirtieth of the ruling cycle time then an error of lead of one-thirtieth of the groove spacing will just halt the blank carriage. Clearly, this is unacceptable since it reintroduces all the problems of slip stick and continuous control at near-zero velocity. In order to avoid this condition, it is necessary to ensure that mechanical errors are less than about one-fiftieth of the groove spacing. This tolerance does not differ greatly from that on groove position derived from the need to reduce the scattering of light through random errors in groove position and emphasizes the need to maintain high standard of workmanship even when interferometric control is applied. The same argument also leads us to expect that for a typical engine with continuous servo-controlled motion of the blank there will be significant reductions in the level of ghost intensities, but that any improvement to the level of stray light (grass) will be less dramatic.

☐   The diamond carriage

In all engines built so far the servo loop is not completely closed. It is the position of the diamond relative to the blank which is to be controlled, but in practice it is the position of the blank relative to some fixed mirror which is measured (and in most cases controlled) while the position of the diamond relative to this reference is determined mechanically. Particular care, therefore, is required in the design of the diamond carriage. It must allow the diamond to reciprocate as freely as possible in one direction and allow it to be raised and lowered accurately and smoothly but allow no movement whatever in the orthogonal direction. Since it is called upon to move at relatively high speeds, it should be as light as possible in order to reduce the out-of-balance forces on the engine, but at the same time it should be as rigid as possible in order to prevent sideways movement of the diamond tip. The guideways of the ruling carriage are the components of the engine which are most susceptible to the effects of wear and of variations of thickness of oil films, so that the materials of the slideways and pads must be chosen with great care. Wear may be reduced by ensuring that the weight of the carriage is borne by slave slide-ways, but since during the course of ruling a grating the diamond travels perhaps three times the total groove length, the rate of wear must still be kept very low and the engine designed to minimize its effects. This is an area in which materials technology has made a useful contribution in that better lubricants are now available and so are materials which combine low friction with very low rates of wear. For example, the reciprocating blank on the Strong engine described in 1951 was guided by steel tapered sleeves riding along cylindrical steel guide shafts lubricated with oil. On the MIT "C" engine guidance was provided by a Rulon shoe in contact with a glass optical flat without lubrication. On Horsfield's hydraulic engine the diamond carriage rode on five sapphire pads bearing on two flat steel bars. On the recent NPL engine both support and guidance are provided by pads of graphite-filled PTFE riding without lubrication on optically flat ceramic ways, a combination which exhibits an extraordinarily low rate of wear.

☐   Environmental control

One of the difficulties in ruling gratings is due to the fact that it takes a long time and that the accuracy of ruling must be maintained without a break and without drifting for perhaps one or two weeks. Various aspects of the environment can affect the position of the grooves and must be kept constant or compensated for in some way. Changes in temperature will give rise to expansion and contraction and must be controlled quite closely. For example, even if all the grooves could be ruled at precisely the correct spacing, there would still be errors in the grating if the blank were to expand and contract during ruling. A more serious effect of variations of temperature is the warping and bending of various parts of the engine caused by differential expansion of different materials or by temperature gradients. Not all parts of the engine will be equally susceptible to variations in temperature. For example, if one side of the blank carriage expands with respect to the other, then the blank will move around an arc of a circle and unless this is corrected (as it is in some cases by an interferometer which servo controls the orientation of the blank) it will give rise to fan error. On the other hand, if the length of the push rod driving the diamond carriage varies in length, then this will have the effect of moving the groove slightly along its own length and this will not be serious. Again, a very small distortion in the "tracelet" holding the ruling diamond can produce catastrophic variations in the position of the diamond tip with respect to the reference mirror for the servo which controls the blank position.

Evidently the temperature must be kept constant, but it is not simply a question of maintaining the ambient temperature because the engine itself inevitably generates some heat due to friction on guideways and motors and light sources. In practice it is usual to control the temperature in various stages. First the room in which the engine is situated is controlled perhaps to $0.5°C$ and then the engine is placed in a thermally insulated box in which the temperature is maintained slightly above ambient and constant to, say, $0.1°C$. Certain particularly sensitive parts of the engine may then be controlled independently to even tighter tolerances. In practice it is very difficult to maintain all parts of the engine at the same temperature if only because different parts have different thermal capacities and therefore respond at different rates to any changes or corrections that may be applied. In order to overcome this problem, some engines, those of Harrison and Stroke at MIT and of Yamada and Harada (1966) in Japan, were partially immersed in a bath of oil.

The control of humidity is also an important feature of maintaining the precision of a ruling engine. Many materials absorb moisture and this changes their physical properties. Oil films may, for example, change their thickness and viscosity and some glues and cements expand when they take up moisture and can exert considerable forces to distort or move the components which they are holding.

It is not usually possible to control barometric pressure and this will affect the engine. The most serious effect is that it causes the wavelength of light to change, and since spacing of the grooves is often based upon the wavelength of light, variations in pressure, unless corrected, will give rise to corresponding changes in groove spacing. There is sometimes an obvious correlation between errors in the diffracted wavefront and variations in temperature, pressure or humidity. Figure 4.7, for example, shows a wavefront interferogram from a grating and a record of the humidity for the period in which it was ruled. This emphasizes that a grating is in effect a record of all the factors influencing the ruling process.

In order to maintain an accuracy of groove position of, say, 25 nm over a grating width of 250 mm, we require the wavelength

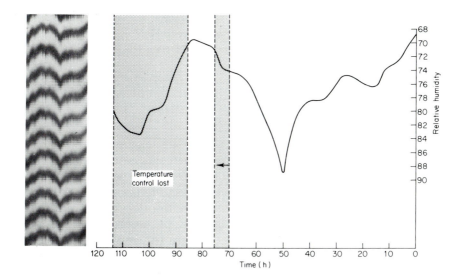

Figure 4.7 The wavefront interferogram of a grating ruled on an engine which was sensitive to humidity compared with the record of the humidity during the ruling process.

to be constant to one part in $(250\,\text{mm}/25\,\text{nm})$, i.e. one part in 10^7. Unfortunately it is the frequency of a light source that can be stabilized rather than the wavelength, and the wavelength is inversely proportional to the refractive index. The refractive index of air varies with pressure according to the expression:

$$\mu - 1 \;=\; 293 \times 10^{-6} P$$

where P is the pressure measured in atmospheres. So, in order to maintain the wavelength sufficiently constant, the barometric pressure must remain constant to approximately 1/3000 of an atmosphere. This cannot be achieved in practice so it is necessary instead to compensate for pressure variation. One particularly simple technique due to Babcock (1962) at Mount Wilson is to use an aneroid barometer to rotate a plane parallel plate in one of the arms of the control interferometer in such a way that it introduces just sufficient path difference to compensate for the change in wavelength. Harrison introduced the correction into the system by means of a differential gear box which generated his reference signal. In the polarizing Michelson interferometer the correction may be introduced by means of a half-wave retardation plate which can be adjusted to introduce any amount of rotation of the plane of polarization of the output beam.

☐   Vibration isolation

Vibrations, which may be generated by the engine or transmitted through the ground, produce relative movements between component parts of a ruling engine which may be too rapid to be corrected fully by servo systems. The most important consideration is the removal of unwanted relative movement between the diamond tip and the grating blank. Vibration amplitudes will be reduced if the engine components are designed to have a high stiffness-to-weight ratio, for the resulting high resonance frequencies of these components will then be remote from the exciting frequencies in the ground. The ground movements under consideration occupy the range 0.25–200 Hz. As a further precautionary measure many ruling engines are supported on anti-vibration mountings, though many others are operated successfully without isolation.

If the mounting is to have a low transmission for vibrations in the important range from 10 to 100 Hz, where the velocity of the ground movements may be large, then its resonance frequencies in the linear modes should be as low as practicable; the natural frequencies of the mounting must not be greater than half the lowest frequency against which isolation is desired. Values between 1 and 3 Hz are suitable, since frequencies below 1 Hz would be uncomfortably close to the fundamental or second harmonic of the ruling frequency. Older designs of anti-vibration mounting using layers of sand or cork have a performance which is difficult to predict at the design stage, and do not give low resonance frequencies. For given natural frequencies in the rotational and linear modes, the required spring stiffnesses for the anti-vibration mounting are directly proportional to the mass supported. Air springs provide a convenient, cheap and compact means of combining high load capacity with low natural frequencies. In order to avoid objectionable tilts arising from translation of heavy components (such as the grating blank), the suspended mass must be heavy (several tonnes) unless servo levelling is employed for the air springs. The consequent inertia of a heavy mounting block is also beneficial in minimizing the effects of vibration generated in the engine itself.

□   Ruling engines — summary

The ruling of diffraction gratings is a task which has occupied the time and attention of many able scientists and engineers for about a hundred years. Some have embarked upon this work because they needed the gratings in order to pursue other work, others have seen the ruling of a "perfect" gratings as the supreme challenge in precision engineering. All would agree that it is not a task to be taken lightly, that the chances of failure are far greater than the chances of success, and that it is a task which demands unlimited reserves of patience. However, without this work progress in spectroscopy would have been severely limited both by the lack of suitable materials for the manufacture of prisms and by their own physical limitations. It is not the purpose of the present section to review the whole subject of ruled gratings, but rather to acquaint the reader with the sort of problems that occur and with some of the solutions that have been found. In most cases there is no obvious best solution and the design of any ruling engine is a compromise between a number of conflicting requirements. Just what form this compromise takes will depend partly upon the personal preferences of the designer, partly upon the constraints within which he works (limitations of cost and the availability of particular skills and materials) and partly upon the type of grating which is required and the use to which it is to be put.

□   Interference gratings

We have seen that the difficulties in ruling gratings are immense and that even with the resources of modern technology the construction and use of a ruling engine is a laborious task with a high risk of failure. This has led many people to consider alternative methods for making gratings. As early as 1872 Lord Rayleigh considered the possibility of the photographic reduction of a large drawing, but concluded that this would be impractical because the original would have to be drawn, as he put it, on the wall of a cathedral. For example, if it were convenient to draw lines at a spacing of 5 mm then a grating that would reduce to 100 mm square with 600 grooves mm^{-1} would occupy an area 300 m square. Quite apart from the difficulty of drawing it, no lens would ever cope with the task of reducing this accurately.[†]

Michelson suggested in 1927 that it might be possible to generate a grating by photographing interference fringes. He proposed a system similar to Wiener's experiment in which standing waves are formed by reflecting light at normal incidence from a mirror and are recorded on a photographic plate which is inclined to the mirror. The width of the grating would then be limited only by the "homogeneity" of the light source, or to use the more modern term, its coherence length. In principle it should be possible to produce a grating just wide enough to resolve the linewidth of the source used to make it. Unfortunately Michelson did not have a suitable source, or suitable materials to try this, but, as with his suggestion of the interferometric control of ruling engines, his ideas were taken up later when such sources and material became available.

In the 1950s J. M. Burch and his colleagues at the National Physical Laboratory (Burch 1960, Burch and Palmer 1961) developed techniques for making gratings by recording interference fringes on photographic plates. Originally this work was directed towards a study of the dimensional stability of photographic emulsions, but gratings were also made for spectroscopic and metrological applications. By bleaching the photographic emulsion it was possible to induce a modulation in the surface and thus to make not only amplitude gratings but also phase gratings with a roughly sinusoidal groove profile. Two types of interferometer were built, one for small fine gratings (up to 2400 grooves mm^{-1}), and one for coarse gratings of large area (e.g. 70 grooves mm^{-1}, 450 mm square), and these are shown in Figure 4.9 (h and i). Both had high luminosity and used as a light source a single-isotope ^{198}Hg lamp. The output was rather low so that exposure times were often of the order of several hours. Unfortunately, the inevitable granularity of photographic emulsions (the grain size of MR plates is about 0.5 μm) gave rise to a level of diffuse scattering from the grating which rendered it unacceptable for the majority of spectroscopic applications. Other materials such as dichromated gelatine and photoresist (e.g. fish glue) were tried, but as they were less sensitive than photographic emulsion they demanded impracticable exposure times.

The problems of long exposure were largely overcome with the

[†] In fact it is now possible to use accurate step and repeat cameras that have been developed for the microcircuit industry to generate reasonable areas of somewhat coarser gratings from a drawing of more manageable dimensions. Such gratings are made for metrological purposes and are discussed in Chapter 9.

invention of the laser and in particular the development of high-powered argon ion lasers in the mid 1960s. The advantages of a laser are that it can generate a beam of light that is accurately mono-chromatic (so that it has a long coherence length) and since all the light is concentrated into a narrow, well defined beam, the power density is much greater than for any incandescent source. With a laser it was possible to use slower but less grainy photosensitive materials and in 1967 both Labeyrie in France and Rudolph and Schmahl in Germany produced in photoresist the first such gratings of a quality which was suitable for spectroscopy. Gratings made in this way became known as "holographic" gratings largely because holography had also become a practical proposition with the invention of the laser and at the time it was attracting a great deal of attention. The justification for the use of the term "holographic" is rather tenous and "interferographic" would be more accurate (interferometric would be wrong since this implies measurement) but for simplicity we shall adopt the term "interference gratings". The essential feature of holography is that a hologram records information about a wavefront in such a way that it can be recon-structed when the hologram is illuminated, and that it is usually made by recording the interference pattern between the original wavefront and a reference wavefront. By analogy a grating consisting of straight parallel equispaced grooves could be regarded as a holo-gram of a point source at infinity and any interferogram could be regarded as a hologram of something. Those working in the field of holography would regard a grating as a simple form of hologram, while those working in spectroscopy would regard a hologram as a complex form of grating.

The process of making interference gratings consists in its simplest form of coating a suitable blank with photoresist, exposing it to the fringe pattern and developing it. The variation of intensity across the pattern is recorded in the resist as a variation of solubility, so that when it is developed in a suitable solvent, the rate of removal of material varies with exposure and gives rise to a corrugation in the surface of the resist. The grating may then be used in trans-mission as it stands or, as is usually the case, it may be vacuum-coated with a reflecting layer of metal and used in reflection.

The substrate is usually glass that has been polished to the required shape within a small fraction of the wavelength of visible light. A thin layer of photoresist which may be between $0.1\,\mu m$ and $1\,\mu m$ thick is applied by spin coating, wherein a small pool of resist is applied to the centre of the blank, which is then rotated at high speed (typically between 1000 and 3000 r.p.m.) and with controlled acceleration. The excess resist is flung off and a thin uniform coating is left on the blank. There are two categories of photoresist, positive and negative. In positive resists the action of light causes long-chain polymer molecules to be broken down and thus become more soluble, whereas in negative resists the reverse is true and it is the unexposed areas that are removed on development. In the application for which photoresists were developed material is either removed completely or left intact, in order to produce a mask which will protect the substrate from chemical etchants. In the present case the resist is subjected to a continuous range of exposures; it is not usually developed right through and the surface remains covered. Although negative resists have been used in this way it is most usual to use a positive resist, partly because they are developed in aqueous solutions which are more convenient than the organic solvents required for most negative resists and partly because with a negative resist is it necessary to ensure that the material in contact with the substrate is adequately exposed in order to be sure that it will adhere to the surface.

The sensitivity of most photoresists is strongly dependent upon wavelength and falls dramatically for wavelengths greater than about 500 nm. Exposures that are required are of the order of 250 mJ cm^{-2}, which is about 1000 times greater than that required for most high-resolution photographic emulsions and 10^5 times that required for modern holographic emulsions. The response is not completely linear and we would therefore not expect the shape of the grooves to be perfectly sinusoidal when exposed to a perfectly sinusoidal variation of intensity. The lack of sensitivity and the need for short-wavelength radiation effectively dictates the type of laser that must be used. Argon ion, krypton ion and helium–cadmium lasers are all suitable as these generate a range of wavelengths in the blue and near ultraviolet region of the spectrum.

If two coherent beams of light intersect at an angle 2θ, as shown in Figure 4.8, they will generate within the volume common to both beams interference fringes with a spacing x given by

$$x = \frac{\lambda}{2 \sin \theta}$$

If a blank coated with photosensitive material is inserted in this region at an angle δ to the bisector of the beams, then fringes will be recorded with a spacing d given by

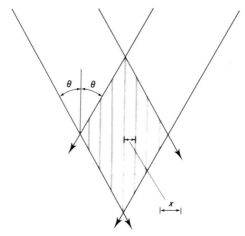

Figure 4.8 The generation of interference fringes at the intersection of two coherent beams of light.

$$d = \frac{\lambda}{2 \sin \theta \sin \delta}$$

The finest spacing that can be obtained is $\lambda/2$ which corresponds to $\delta = 0$ and $\theta = 90°$. This cannot be achieved in practice, since it requires both beams to be incident along the surface of the blank, but a value of $\theta = 60°$ yields a spacing of only 0.6λ. It follows that the minimum spacing that can be obtained with the argon (458 nm) line is $0.28\,\mu$m or about 3500 grooves mm^{-1}; with the krypton (351 nm) line it is 0.21μm or 4700 grooves mm^{-1}; and with the frequency-doubled argon (257 nm) line it is $0.15\,\mu$m or 6600 grooves mm^{-1}. Finer pitches still can be obtained by causing the interference to take place in a dense medium. This is equivalent to reducing the wavelength by a factor equal to the refractive index and Figure 4.9(j) shows a system used by Labeyrie to obtain pitches of over 6000 grooves mm^{-1} with light of a wavelength of 488 nm.

With interference gratings it is (within limits) possible to choose any pitch and the exposure time required is almost independent of this. On the other hand in the ruling process the time taken is proportional to the pitch and the difficulties increase even faster. Diamond wear usually limits the total groove length that can be achieved, whereas interference techniques are not limited in this way and it is therefore possible to make very large gratings with very fine pitch.

☐   Interferometers

There is a wide variety of optical systems that may be used to bring together two beams of light in such a way that they will generate interference fringes; some of these are shown diagrammatically in Figure 4.9.

In all cases the two interfering beams are generated from the same light source in order to ensure coherence. The two beams may be obtained either by the division of amplitude, that is with a beam splitter which reflects a fraction of the intensity of the whole beam and transmits what is left after absorption, or they may be obtained by the division of wavefront, in which case one half of the wavefront pursues a different optical path to the other half.

Figure 4.9(a) depicts the system originally proposed by Michelson. It has the virtue of extreme simplicity, but it does mean that the light has to pass through the back of the blank which can be inconvenient unless special precautions are taken to reduce the effects of scattering and of multiple reflections within the blank. We shall see later that it generates blazed gratings.

The system of Figure 4.9(b) is again very simple and can be regarded as an extension of Michelson's arrangement. It requires the substrate to be metallized in order to reflect the beam and in fact produces a rather complex interference fringe pattern (Boivin 1973) in which there are two sets of fringes, one perpendicular to the blank and the other parallel to it. In order that the surface of the resist should be given the maximum exposure, it is necessary to ensure that the thickness of the resist layer is such that the surface corresponds to an antinode of the fringe system parallel to the blank.

Figure 4.9(c) shows a Fresnel prism in which two halves of a collimated beam are refracted across each other's path and fringes are produced. This is a very simple and stable interferometer, but for a given wavelength of light the spacing of the fringes is fixed. It also requires a prism which is larger than the grating to be made, so it is not practical for the manufacture of large gratings, particularly if these have to be of high quality. Any inhomogeneity of the glass will produce local deviations of the beams and hence local variations of grating pitch and it is extremely difficult to find large pieces of glass of suitable homogeneity.

The arrangement shown in Figure 4.9(d) was proposed by Labeyrie. It has the advantage that there are no transmitting optical components after the pinholes and that each beam only encounters one optical surface. However, in order to achieve perfectly collimated

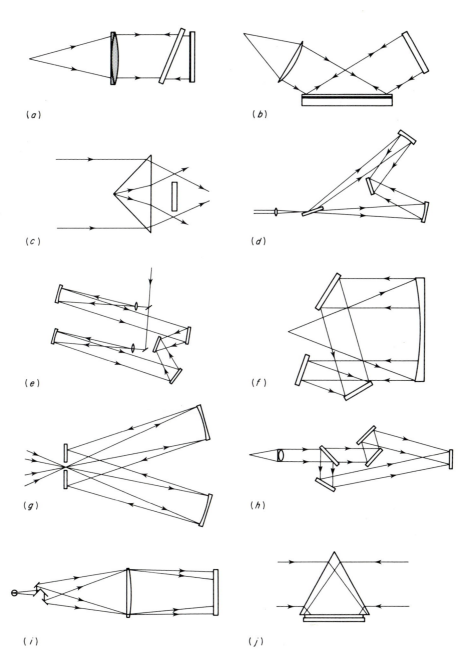

Figure 4.9 Various forms of interferometer for the production of interference gratings.

beams it requires the use of two off-axis paraboloidal mirrors. These are difficult to make and because they are not self-generating surfaces the degree of surface polish is often inferior to that of a spherical or a plane mirror. This gives rise to scattering of the incident light which is recorded on the grating. The alternative to using aspherical mirrors is to use spheres of a sufficiently small numerical aperture that the aberrations of the collimated wavefronts may be neglected. This leads to rather long focal lengths and demands the use of a large optical table particularly for large gratings. Such a system is rather sensitive to draughts and thermal gradients.

Some of these problems are overcome in the system of Figure 4.9(e) because the two collimating mirrors are used in the same orientation and if they are identical, for example perfect spheres, then the aberrations from the two beams will be identical and with care they can be arranged so that they cancel each other out.

Figures 4.9(f) and (g) are two systems which have the advantage of using the collimating mirrors on-axis. That of (f) makes efficient use of the available light but has the disadvantage that it requires a mirror which is more than twice the width of the grating to be made. That of (g) has the disadvantage that the light passes through a hole in the grating blank itself. In many cases this would be quite unacceptable, but for very large gratings used in astronomy this may not matter because it frequently happens that the central region of the grating is masked by, for example, the secondary mirror of the telescope.

System 4.9(h) is comparatively complex in that it requires three plane mirrors and a beam splitter in addition to the collimator. It has the advantage that the aberrations of the collimated beam tend to cancel since they affect both beams equally. It requires the plane mirrors and beam splitter to be of the highest optical quality, but will perform satisfactorily with a comparatively crude collimator such as a spherical mirror used off-axis or a single planoconvex lens. In practice it is usually easier to make good optical flats than it is to design and make collimators of equivalent optical quality. The system has the disadvantage that it is sensitive to inhomogeneities in the beam splitter, because only one beam passes through it, and for this reason it is unlikely to be suitable for the manufacture of high-quality gratings wider than about 150 mm.

Figure 4.9(i) is an equivalent system for making coarse gratings. It has the same immunity to aberrations of the collimator but saves a great deal of space by placing the beam splitting arrangement behind the collimator. It could be considered as a rather more sophisticated form of Young's slit experiment. In this case the size is limited by the aperture of the collimator and not by the beam splitter and coarse gratings up to 15 in.[2] were made on such a system by Burch in 1960.

Figure 4.9(j) illustrates the method used by Labeyrie to generate gratings of very fine pitch by recording the fringes inside an optically dense medium.

In order to obtain the very best possible quality of diffracted wavefront, Rudolph and Schmahl (1970a, b) developed an elegant technique in which the interfering wavefronts were generated holographically. By making two holograms of a single good quality plane wavefront they were able to generate two good wavefronts and to cause them to interfere in such a way that any residual aberrations cancelled. The method demands long exposure times, because much light is lost due to the inefficiency of the holograms, but using this technique Rudolph and Schmahl produced (genuinely holographic) gratings with excellent wavefronts.

Some of the interferometers shown in Figure 4.9 have the feature that most of the aberrations are common to both wavefronts and their effects tend to cancel. This cancellation can be thought of in two ways, which are depicted in Figure 4.10.

The position of any given fringe depends upon the difference in path length from that point to the source via the path of the two beams. If this difference is an even number of half-wavelengths then the fringe is bright, and if it is an odd number of half-wavelengths then the fringe is dark. An aberration of a wavefront may be regarded as the local addition of an extra optical path so that one region of the beam lags behind the rest. If the same extra optical path is introduced into each beam, as in Figure 4.10(a), then the position of the fringes will not be affected; all that will happen is that the intensity reaches its maximum at a slightly different time.

The alternative view of the cancellation of aberrations considers the wavefront as distorted locally so that its normal and hence direction of propagation are different from those of the ideal wavefront. If both interfering wavefronts have the same aberration, as in Figure 4.10(b), then the angle between their normals is the same and the spacing of the fringes is the same. In effect the blank is inclined to these fringes at a small angle, but since the pitch of the grating that is recorded varies with the cosine of this angle the variation may be neglected for small angles.

In this connection it is interesting to note that some defect of focus in the two beams may be tolerated if both defects are in the

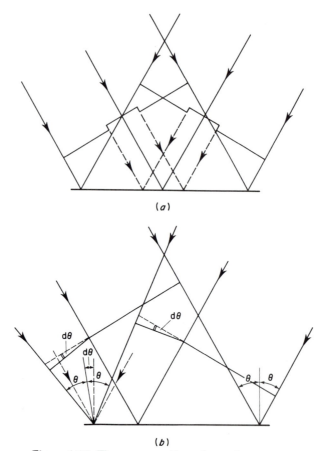

Figure 4.10 The compensation of wavefront errors.

same direction. For example, if both wavefronts are slightly concave the above argument applies and the pitch varies with the cosine of the angle between the actual wavefront and the notional plane wavefront. If one wavefront is concave and the other convex then the blank is still normal to the bisector of the beams, but the pitch of the *fringes* changes according to the sine of the angle between them. At the angles commonly used this varies much more rapidly than the cosine.

☐   Setting up the interferometer

It is often necessary to record a grating with a particular value of groove density and in order that such a grating shall be compatible with an existing instrument the tolerance on pitch is often quite severe. One might encounter, for example, a need for the pitch to be accurate to $0.1 \, \text{grooves} \, \text{mm}^{-1}$ (one $\text{groove} \, \text{cm}^{-1}$) on a 2000 $\text{groove} \, \text{mm}^{-1}$ grating. Therefore the sine of the angle between the grooves must be accurate to 5 parts in 10^5 or the angle set typically to 10 seconds of arc. This may be achieved by means of a mirror mounted on an accurate rotary table in the position of the grating blank, but there is a far simpler technique if a grating of the required pitch (or a sub-multiple of it) is already available. The grating is then set up in the position of the blank, preferably with some means of rotation in its own plane so that it may be adjusted with its grooves parallel to the fringe pattern, as shown in Figure 4.11.

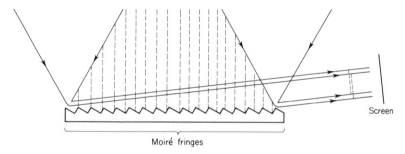

Moiré fringes

Figure 4.11 The adjustment of the spacing of the fringes by comparison with a grating.

One may then either view on the surface of the grating the moiré pattern between the grating itself and the "stripey light" of the fringe pattern, or one can observe on a screen the interference between coincident wavefronts diffracted from the two incident beams by the grating. The angles of the incident beams are then adjusted until this fringe pattern is "fluffed out" (i.e. to as near a uniform an intensity distribution as possible), thereby giving an automatic adjustment of the fringe spacing. In this way it is possible to reproduce a given pitch to better than one groove per grating. This is far more than is necessary for any spectroscopic application, since the error in position of any given spectral component would be less than the resolution width of the grating.

☐   Spectral performance

Granted that it is now possible to make gratings by recording interference fringes in photoresist, we must now ask how they perform
in practice and how they compare with their ruled counterparts.
We shall defer until the next chapter a full discussion of the properties of gratings and how to measure and define them. For the
present, let it suffice to say that there are three properties which
are of interest to the spectroscopist: the resolving power, the spectral
purity and the efficiency. The extent to which any particular feature
is important will depend upon the use to which the grating is to
be put, but in all cases a grating has to reach certain standards
before it may be considered useful.

As we saw in Chapter 2, the theoretical resolving power is primarily a function of the size of the grating, although the pitch and order
in which it is used are also relevant. In principle the size of an interference grating is limited by the coherence length of the light source,
so that it is possible to make a grating which will just resolve the
spectral width of the source. The coherence length of an argon or
krypton laser operating with an intracavity etalon is several metres,
so that the limit of size is determined in practice by economic
constraints rather than by the laws of optics. The manufacture and
handling of large optical components is both difficult and costly
and is generally only justified in the case of gratings used for astronomy. In this case it is usually the light-gathering power rather than
resolving power which is of prime importance. Gratings with diag-
onals or diameters of the order of 500 mm have been both ruled and
produced by interference techniques. However, because of the limitations of diamond wear, the ruled gratings were echelles to be used
in high diffracted orders and the interference gratings were of fine
pitch for use in first order. An interference grating of the same
pitch as the echelle would have too low an efficiency to be useful,
so that in the case of large gratings the two techniques do not
compete.

If we assume that a grating is to be used in conjunction with a
perfect optical system, then the fraction of the theoretical resolution
that is achieved will be determined by the quality of the spectral
image and this in turn depends upon the aberrations of the diffracted
wavefront. The Rayleigh tolerance states that there will be no appreciable loss of resolution if the wavefront does not depart from a
plane or spherical surface by more than one quarter of a wavelength.
This is a useful general rule but there are some exceptions and, for

example, if a grating is working in an instrument with slits then some vertical astigmatism (i.e. a cylindrical wavefront) may be tolerated. It is usually possible to apply the photoresist with sufficient uniformity to meet this tolerance so that any aberrations are generally those introduced by the interferometer. Using good optical components and by careful design of the interferometer it is possible to make interference gratings which are diffraction-limited in the visible and as good as ruled gratings. If resolving power or wavefront flatness are of prime importance, the holographic technique of Rudolph and Schmahl will provide the best possible results. In general we may say that both ruled and interference techniques are capable of producing gratings of adequate resolving power, so from this point of view alone neither technique offers any particular advantage.

The parameter for which the interference grating does offer a significant advantage is the "spectral purity". Despite the best efforts of the ruling engineers, it has not been possible to reduce mechanical errors in the position of the grooves to the state where ghosts and grass are completely eliminated. In an interference grating the positions of the grooves are dictated by the conditions of interference, so there are no random or short-term periodic errors and ghosts and grass are completely absent. It is possible to generate ghosts with an interference grating if, for example, there are two or more wavelengths coming from the laser (as might happen with the Ar and Kr UV lines) or if a secondary reflection from some optical component (such as a beam splitter) generates an extra image of the source. However, with care these can be eliminated and ghosts can be completely removed from the spectrum of interference gratings.

There are often sources of stray light which are peculiar to interference gratings and again it demands a certain amount of care to avoid or to reduce them. For example, when the photoresist is spun onto the blank any dust on the surface will obstruct the flow of the resist and give rise to a radial streak or "comet mark" where the resist is of a different thickness. This effect is shown in Figure 4.12 and will generate stray light in the form of a halo around the main diffracted order in the spectrum. Another possible source of stray light in interference gratings is light which is scattered from the optical components and mounts of the interferometer. There is a danger that any such light which falls upon the blank during recording will in effect form a hologram of the apparatus and that when the grating is used the reconstruction from the hologram will be detected as a form of stray light. As with ghosts, both of these sources of

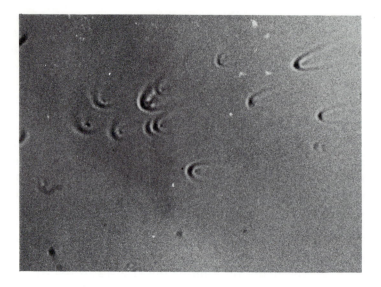

Figure 4.12 Imperfections in a photoresist coating caused by dust on the substrate.

scattering can be eliminated provided that sufficient care is taken over the preparation of the blanks and the setting up of the interferometer.

We shall see in the following chapter that the measurement of stray light is not straightforward and a comparison of the scattering from different types of grating must be made with extreme caution. However, if one illuminates both a ruled grating and an interference grating with an unexpanded laser beam and observes on a screen the level of light between the diffracted orders, then the difference between the two types is dramatic. This is shown in Figure 4.13, which shows quite clearly that in the "spectrum" from a ruled grating there is a line of ghosts and grass joining the diffracted orders, whereas for an interference grating this is absent. Of course, whether the stray light which we observed from a ruled grating actually *matters* depends upon the application for which it is intended. Indeed in some instruments, stray light from other components makes a significant contribution to the overall level of stray light. Despite these observations it is generally agreed that interference gratings exhibit a significantly higher degree of spectral purity than do ruled gratings.

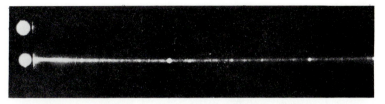

Figure 4.13 The comparison of stray light from an interference grating (top) and a ruled grating (bottom).

The third feature of grating performance, one which is always of interest even if it is not of prime concern, is the efficiency. There may be applications where the level of light is so high that one can afford to lose most of it with an inefficient grating provided that the signal-to-noise ratio or the resolution is adequate. However, such applications are not common and in general a grating will only be acceptable if a significant proportion of the incident light is diffracted into the required order. In this respect interference gratings differ from ruled gratings because the interference technique affords less control over the groove profile. Nevertheless, for a given application the choice between a ruled and an interference grating is not always straightforward and may depend, for example, upon whether one has the freedom to select the pitch of the grating.

The simplest form of interference grating, recorded on any one of the interferometers shown in Figure 4.9† will have a symmetrical groove profile. The two wavefronts impinge upon the blank at equal angles and generate a sinusoidal variation of intensity across the fringe pattern. This is recorded in the photoresist and, although non-linearities of the resist may distort the sine wave, the symmetry remains and the groove profiles will resemble those shown in Figure 4.14, which contrast with the saw-tooth profile usually obtained with ruled gratings.

In assessing the efficiency of interference gratings as a whole we must ask two questions. Firstly, "to what extent is it possible to use gratings which have a symmetrical groove profile?" and secondly "to what extent is it possible to generate with interference techniques gratings which have an asymmetrical or blazed groove profile?". With a symmetrical profile and light at normal incidence it is evident that the efficiency in any order cannot exceed 50% because, by symmetry, the positive and negative orders are equal; in practice the efficiency seldom exceeds 33%. However, in most cases the grating is not used in normal incidence, so that positive and negative orders

† Apart from 4.9(*a*).

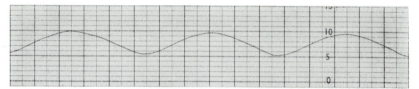

Figure 4.14 A sinusoidal groove profile generated in photoresist.

are diffracted at different angles. The argument of symmetry thus
no longer applies and there is no *a priori* reason why the efficiency
of a grating with a symmetrical profile should not exceed 50%.
Indeed, if the pitch of the grating is sufficiently fine, then it may
be that the diffracted order of opposite sign to the one in use cannot
exist. It is evanescent or "diffracted below the horizon" because the
sine of the angle of diffraction exceeds unity. In practice it has been
found that it is possible with a symmetrical groove profile to achieve
efficiencies that are just as great as those obtained with a blazed
profile, provided that the grating is used in a configuration that
allows the propagation of only one diffracted order (in addition of
course to the zero order). In this case the incident light is distributed
between the diffracted order and the zero order in a proportion
which depends primarily upon the groove depth. Theoretical studies
have shown that under these circumstances the distribution depends
remarkably little on the shape of the groove profile and this has
been borne out by experiment. Figure 4.15 for example shows
efficiency curves for a ruled grating and an interference grating of
1800 grooves mm^{-1}.

One can think of any groove profile as a Fourier synthesis of a
series of sinusoidal profiles with a harmonic series of spatial frequencies. In the single-diffracted-order configuration a grating with

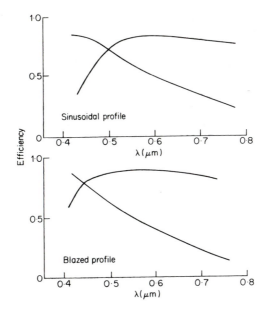

Figure 4.15 Comparison of the measured efficiency curves for 1800 grooves mm^{-1} gratings with (*a*) sinusoidal and (*b*) blazed groove profiles.

a groove density double that of the fundamental grating would sustain no diffracted orders, so that this and higher harmonics might be expected to contribute little to the far field diffraction pattern. Although this should not be taken as a rigorous argument, it does offer a plausible explanation of the insensitivity of efficiency to groove profile and of the observed fact that high diffraction efficiencies may be obtained with gratings with a symmetrical groove profile. This argument has been developed more rigorously by Breidne and Maystre (1980).

However, it is not sufficient simply to point to efficiency curves such as those of Figure 4.15 and to conclude that interference gratings are as efficient as ruled gratings. They may be, but the condition of a single diffracted order imposes some severe restrictions on the mounting and use of a grating and needs to be considered in rather more detail. We start, as ever, from the grating equation:

$$d(\sin \alpha - \sin \beta) = m\lambda$$

The condition that an order m is just about to disappear into the surface of grating is that β equals $\pm 90°$, so we may now write

$$d(\sin \alpha \mp 1) = m\lambda$$

or

$$\alpha = \sin^{-1}\left(\frac{m\lambda}{d} \pm 1\right)$$

We may now plot α as a function of λ/d for various values of m and this will show us how many diffracted orders are able to propagate for a given combination of λ, d and α. In Figure 4.16(a) we see that for the ± 1 order to be unique, λ/d must exceed 0.66 at an angle of incidence of 19.5°. In order to put this diagram into perspective it is useful to add some further features. With the exception of echelles, it is generally inconvenient to work with very high angles of incidence and diffraction because the effective aperture of the grating becomes very small. If we assume (quite arbitrarily) that the limiting angle of incidence and diffraction shall be 60°, then we may plot this condition on the diagram and mask off those areas which it excludes. This is shown in Figure 4.16(b). It is also possible to draw the curves corresponding to different spectrometer mountings. For example, in a spectrograph the angle of incidence is constant. For a first-order Littrow mounting $\alpha = -\beta$ and $\sin \alpha = \lambda/2d$, whereas for an Ebert or Czerny–Turner mounting there is a fixed angle between α and β, so the line is displaced slightly.

From this diagram we are now able to learn something of the consequences of using a grating in the single-diffracted-order condition. We are restricted to the approximately kite-shaped area of the diagram. The region extends from $\lambda/d = 0.66$ to $\lambda/d = 1.7$, so that we have available a little over one octave of useful spectrum, provided that we use the grating somewhere near the Littrow mounting. The wavelength range to which this corresponds can be chosen at will, provided that we can select a grating with a suitable pitch. In short, it is possible to obtain reasonably high efficiencies for wavelengths that are about the same as the groove spacing.

We saw in Chapter 2 that the facet angle of a blazed grating was the same as the Littrow angle at the blaze wavelength. The shortest wavelength in the "useful region" is that given by the intersection of the $m = -1$ and $m = +2$ lines and corresponds to an angle of incidence of about 19.3° ($\sin^{-1}$ ($\frac{1}{3}$)), but in practice one would choose a blaze wavelength somewhat further into the desired spectral region. This then implies that the region of a single diffracted order corresponds to that for which one would choose a *first-order* blazed grating with a facet angle of, say, 21° or more. In other words,

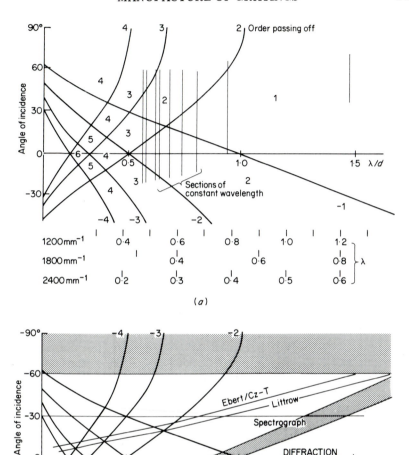

Figure 4.16 (a) Graphs of the angle of incidence versus the normalized wavelength for which various diffracted orders are able to propagate; (b) the relationship between the "passing off" of orders and common instrumental profiles.

rather than rule a grating with such a facet angle, one might as well use an interference grating and take advantage of the extra spectral purity and the greater ease of manufacture.

We see from Figure 4.16(b) that the useful spectral range depends on the type of instrument in which the grating is to be used and it is worth noting that the range for a spectrograph in which the angle of incidence is fixed is rather less than that for a monochromator in which the grating rotates. In a spectrograph the whole spectrum is brought to focus simultaneously and one point which is not brought out in the diagram is that the single-order condition implies that the grating is used in a configuration with rather high dispersion. In a monochromator this can be an advantage because for a given resolution it permits the use of wider slits and more light is able to enter the instrument. For a spectrograph, on the other hand, this dispersion may be an embarrassment because the higher the dispersion the greater is the numerical aperture of the optics required to focus the spectrum. According to the diagram, the maximum spectral range corresponds to an angle of incidence of $60°$. In this case the spectrum goes from $\lambda/d \simeq 1$ to $\lambda/d \simeq 1.7$ (rather less than an octave) and is dispersed through an angle of about $53°$. Thus the camera optics (lens or mirror) have to work at $f/1.0$: a very stringent requirement which is unlikely to be met in practice. It follows that gratings with sinusoidal or other symmetrical profiles are unsuitable for use in spectrographs except when a comparatively narrow region of the spectrum is to be studied.

In many cases it is impractical to use a grating in a configuration in which there is only one diffracted order. Therefore, if a symmetrical, unblazed grating is used it will be less efficient than a blazed grating and often this loss is too high a price to pay for the improvement in spectral purity associated with an interference grating. We must ask, therefore, to what extent it is possible to use interferography to generate gratings with a blazed saw-tooth groove profile — in particular with facet angles less than $20°$. Several different approaches have been suggested and all have met with some success, although so far none has solved the problem completely.

The first method is that of Fourier synthesis (Rudolph and Schmahl 1968, Bryngdahl 1970, McPhedran *et al.* 1973). Rather than expose the resist to a simple sine wave fringe pattern, it is exposed to a series of sinusoidal patterns of different spatial frequencies in such a way that they add up to give a Fourier synthesis of the required saw-tooth exposure pattern. In practice only two components of the Fourier series have been considered and in theory this should be sufficient to generate a reasonable approximation to a saw-tooth, as we see from Figure 4.17. Unfortunately this method demands that the interferometer be set up with far greater accuracy than that required in the case of a single-exposure grating. First it demands that the ratio of the spatial frequency of the different fringe patterns must be an exact integral because, if there is any mismatch, moiré fringes will be recorded on the grating. For example, if the first harmonic of a 1200 groove mm^{-1} grating is 2400.1 grooves mm^{-1} rather than 2400 grooves mm^{-1}, then there will be moiré fringes parallel to the grooves at a spacing of 1 per 10 mm. As one scans through a moiré fringe the relative phase of the two components changes so that the blaze would alternate between positive and negative orders with symmetrical regions in between. In order to eliminate these effects, the spatial frequency of the first harmonic must be accurate to 1/10 groove per *total grating width*. In the case of a frequency of 2400 grooves mm^{-1} made with light of wavelength 458 nm for a grating 100 mm wide, this implies that the angle between the beams must be accurate to about 0.06 seconds of arc — that is one tenth of the angular width of a diffraction-limited image formed by the aperture of the grating. For a single-exposure grating the tolerance on pitch may be 50 times less severe, and in order to obtain the desired groove profile it is necessary only to optimize the exposure and development. In a double-exposure grating it is necessary to generate two fringe patterns of which at least one needs to be very accurate. These must then be aligned and adjusted to have the correct relative phase and then the two exposures must be controlled so that each Fourier component has the correct amplitude.

There have been various successful attempts at overcoming these

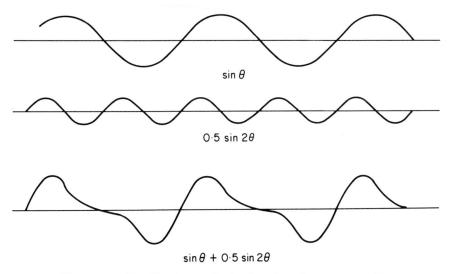

$\sin \theta$

$0{\cdot}5 \sin 2\theta$

$\sin \theta + 0{\cdot}5 \sin 2\theta$

Figure 4.17 The Fourier synthesis of a triangular groove profile.

problems and the method has been shown to work in principle although, at the time of writing, only comparatively small gratings have been produced. One way is to keep the geometry of the inter-ferometer fixed and expose with two wavelengths, one of which is double the other. This is just about feasible using 514.5 nm light from an argon laser and frequency doubling to obtain 257 nm radiation. The photoresist is very insensitive to light of 514.5 nm, so this demands high power and long exposure times. However, it is very sensitive to 257 mm, but the conversion efficiency in frequency doubling is very low so that again long exposure times are required. Problems also arise because although the *frequency* of the light is accurately doubled, the *wavelength* depends upon the refractive index of the air and this is not the same for both frequencies.

Rudolph and Schmahl (Schmahl 1974) overcame many of the problems by using the various diffracted orders from a single grating to generate the beams to produce the required intereference patterns. Their system is shown in Figure 4.18; with it they produced gratings blazed for about 300 nm. The method has the disadvantage that it requires a master grating to be over five times the width of the final grating, although it could be made up of two incoherent halves. It also places severe demands on the flatness of the incident wavefront and the quality of the master grating. The relative phase between the two Fourier components can be controlled by the rotation of a plane parallel glass block inserted into one half of the beam.

An alternative approach has been developed by Johansson and Nilsen in Stockholm, in which they make two quite separate exposures and change the angle of the beams in the interferometer between them. In order to maintain the accuracy of the fringe spacing, they first record around the perimeter of the blank a grating with a spatial frequency equal to half of that required in the final grating. They then use the moiré (as in Figure 4.11) between this and the fringe pattern to align and servo control the interferometer. In this way they have generated gratings with a good approximation to a saw-tooth groove profile and which exhibit high values of efficiency. However, there are severe practical problems in making uniform gratings with apertures much greater than 50 mm. (Breidne *et al.* 1979)

A second method of making blazed gratings directly in photoresist was described by Sheridon in 1968 as a technique for making holograms. Here the photoresist layer is aligned obliquely to the fringe pattern so that there are, within the thickness of the resist, layers which are alternatively soluble and insoluble. After

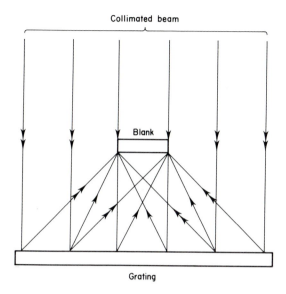

Figure 4.18 The Fourier synthesis of an interference grating using different wavefronts diffracted from a master grating (after Schmahl 1974).

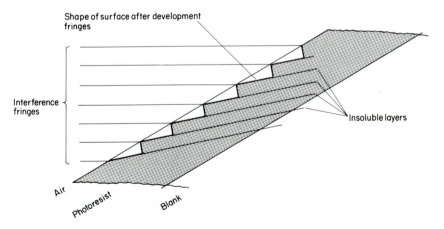

Figure 4.19 The generation of a saw-tooth groove profile in photoresist by inclining the blank to the interference fringes.

development the surface profile is determined by the shape of the insoluble layers near the surface, as shown in Fig. 4.19. In principle it is possible to generate a wide range of groove profiles in this way, but in most cases it is necessary that one of the beams should be incident through the back of the blank. A convenient optical system for producing a restricted range of groove profiles is that originally suggested by Michelson (1915) and is shown in Figure 4.20(a). This entails setting up standing waves by reflecting a collimated beam back down its own path. It is interesting to note that with this system the Littrow blaze wavelength is independent of the angle of inclination of the plate. The facet angle and the pitch both change as this angle varies, but they do so in such a way as to maintain a constant blaze wavelength, as illustrated in Figure 4.20(b).

We would expect on the basis of the simple description of the process that the blaze wavelength should be double the spacing of the fringes *inside the resist layer*. As this spacing is $\lambda/2n$ (where n is the refractive index), the expected blaze wavelength would simply be λ/n. In practice, because there is a sinusoidal rather than a binary

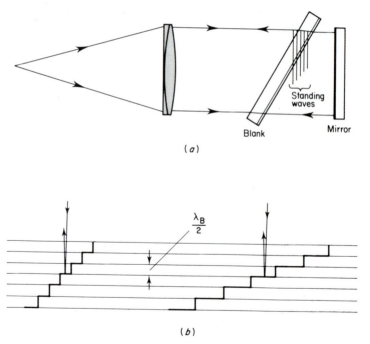

Figure 4.20   (a) A simple apparatus for the production of blazed gratings in photoresist; (b) the invariance of blaze wavelength with grating pitch.

variation of solubility of resist, the facets are rather less steep than would be expected and the blaze wavelengths are correspondingly lower (Hutley 1975). So with present laser wavelengths and photo-resists, the method is best suited to gratings for the ultraviolet spectral region between 300 and 100 nm. However, the quality of the profile is very good (Figure 4.21) and the efficiencies of such gratings are just as high as those of the equivalent ruled gratings. Figure 4.22 shows the distribution of light among the different diffracted orders for such a grating and illustrates just how effectively a triangular groove profile concentrates the diffracted light into a chosen order. Despite the fact that they are blazed for the near ultraviolet, such gratings are frequently used in spectrophotometers covering the whole of the spectral range from ultraviolet to the near infrared. It is an application for which a slightly longer blaze wavelength would have been pre-ferred, but the improvement in stray light compared with a ruled grating compensates for the small loss in efficiency. This technique has the advantage that it requires only a single exposure and the limi-tations of size are no greater than for any other interference grating.

An entirely different approach to the problem of generating triangular groove profiles in interference gratings involves accepting the unblazed profile in photoresist and then changing it by some external means. In this case the prime function of the resist is to record the position of the grooves. In the first technique, which was described by Ayoagi and Namba (1976), a sinusoidal photoresist grating is etched into a layer of poly(methyl methacrylate) (PMMA) by a beam of energetic ions impinging obliquely on the surface as shown in Figure 4.23(a). By choosing the aspect ratio of the sinus-oidal grating and the angle of incidence of the beam, it is possible to generate good saw-tooth groove profiles without the same restric-tions on blaze wavelength that apply to the previous method. A second technique, used by Tsang and Wang (1976), involves making the grating on a single-crystal substrate, developing the resist right through to the substrate so that the resist forms a mask. The grating pattern is then etched into the substrate either chemically or with ion beams and the profile is determined by the crystalline structure of the substrate. This is shown diagrammatically in Figure 4.23(b). A different scheme, which was proposed by Neureuther and Hagouel (1974), uses a laminar grating derived from a photoresist grating to act as a mask for X-ray lithography. The desired profile is obtained by exposing a suitable substrate (coated for example with PMMA) to two X-ray beams which are incident at angles appropriate to give the required facet angles, as in Figure 4.23(c).

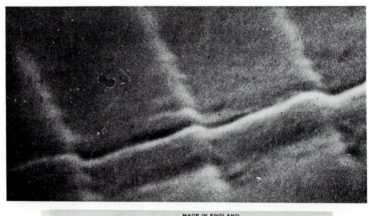

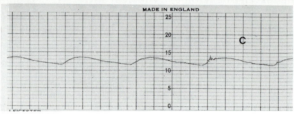

Figure 4.21 Examples of the groove profile of gratings generated according to Figure 4.20.

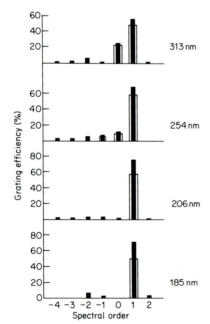

Figure 4.22 The distribution of light among the different orders of diffraction for a grating made according to Figure 4.20; solid lines relative efficiency; open lines, absolute efficiency.

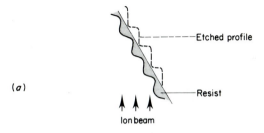

(a)

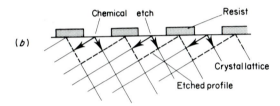

(b)

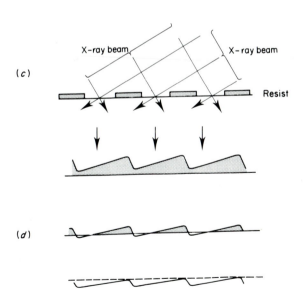

(c)

(d)

Figure 4.23 Various methods of generating a triangular groove profile by etching a photoresist grating.

Finally, it is possible to start with a blazed photoresist grating and change its facet angle by ion etching as depicted in Figure 4.23(d). A beam of ions will remove material at a rate which depends upon the nature and energy of the ions and will vary from one material to another. In general, photoresist and the substrate material will be etched at different rates, so that if etching is continued until the resist is completely removed the depth of the groove is changed by the ratio of the etch rates. Thus if photoresist etches four times as fast as silica, a photoresist grating blazed at 200 nm will be transformed into a silica grating blazed at 50 nm. In principle it should be possible both to increase and to decrease the blaze wavelength, but at the time of writing only shallower gratings have been made successfully (Stuart et al. 1976). The technique significantly extends the range of blaze wavelengths that can be obtained from interference gratings and is especially important for X-ray gratings which are used at grazing incidence. Figure 4.24 shows measured traces of the groove profile of a photoresist grating before and after etching and indicates the extreme shallowness of the grooves that may be obtained and also that they do retain a reasonable saw-tooth shape. It is in fact very difficult to produce gratings as shallow as these, with facet angles down to $0.5°$, by ruling with a diamond.

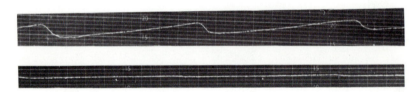

Figure 4.24 The reduction of groove depth by ion etching a blazed photoresist grating on a silica substrate (a) before etching, (b) after etching. The Littrow blaze wavelength is about 35 nm.

Quite apart from the ability to control the blaze wavelength, any technique which results in a grating formed in the substrate without any residual photoresist has the additional advantage of robustness. Not only is it less susceptible to physical damage, but it is also possible to clean off and renew the reflective coating. These considerations are particularly important when the grating is to be used in an ultrahigh vacuum system (when it might have to be baked) or exposed to very high densities of radiation as, for example, in an electron synchrotron, where a normal replica simply boils! We may now summarize the relative merits of ruling gratings and of generating them interferographically. The comparison can be made from two points of view: first from that of the manufacturer and second from that of the spectroscopist. The two are not entirely independent because a grating that is difficult to make, even if its performance is perfect, is of little value if it is not available to spectroscopists. The problem is greatly relieved by the technique of replication in which the grating is copied by casting and the replica rather than the original is used. This means that once a grating has successfully been replicated so that further replicas can be taken from this replica, the availability depends upon the replication process and not on the production of masters.

From the point of view of the manufacturer, the great advantage of the interference technique is that it is simpler and faster. Because it is faster it is not necessary to provide such an elaborate system of environmental control and the ambient conditions need to remain constant in principle only for the few minutes of the exposure rather than for the days or weeks required for ruling. The main disadvantage of the method is that there is less control over the groove profile. Although techniques for generating blazed profiles have been and are still being developed, they are still rather restricted. With a ruling diamond one has some control over both the angle between the facets and the inclination of the facet and this is particularly important when one seeks to eliminate "anomalies" in the efficiency curves.

From the point of view of the spectroscopist, the main advantage of interference gratings lies in their freedom from ghosts and stray light. As far as resolving power is concerned, there is little to choose between the two types so it is probably true to say that most spectroscopists would prefer to use an interference grating, provided that is has suitable efficiency characteristics. It is, therefore, particularly important to understand just when this is likely to be the case. With the present state of the art (1981), the selection rules can be summarized as follows:

(1) If the grating is to be used in first order and the facet angle of the optimum blazed grating would be greater than about 21°, then a simple interference grating with a sinusoidal profile can be used without loss of efficiency.

(1a) In other words, if the spectral range required is an octave or less and high dispersion is acceptable, then a sinusoidal grating with a groove spacing approximately equal to the centre of the wavelength band can be as efficient as a ruled grating.

(2) If the Littrow blaze wavelength required is 250 nm or less, then it is possible to produce a blazed interference grating with as high an efficiency as the best ruled grating independent of pitch.

(3) If the signal-to-noise ratio is of great importance, then it is worth considering the use of a blazed interference grating even though it may not have the optimum blaze wavelength. It may be that the substantial gain in spectral purity will offset any loss of efficiency that is entailed.

(4) The interference technique is well suited to the manufacture of gratings of fine pitch but inappropriate for coarse gratings. There is at present no substitute for a ruled echelle. For any application demanding a pitch in excess of 1200 grooves mm^{-1}, the use of an interference grating should be seriously considered.

(5) The pitch of the best ruled grating for a given application may well be different from that of the optimum interference grating. If an interference grating is being considered as a replacement for a ruled grating in an existing instrument, the pitch may be fixed and it may be impossible to use interference gratings to their best advantage. On the other hand, if one is designing an instrument from the beginning, it may well be possible to benefit from the use of interference gratings.

There is one further advantage of the interference technique and that is its versatility. Gratings can be formed on curved substrates upon which it would be impossible to rule. The subject of concave and other "self-focusing" gratings is dealt with more fully in Chapter 7, but it is worth noting here that although gratings are ruled on concave spherical surfaces and occasionally on toroidal surfaces, the practice is limited to gratings of rather shallow curvature. The interference technique is not limited in this way and the ability to make gratings on, for example, very steeply curved toroids provides the instrument designer with far greater scope.

Not only can gratings be formed on a variety of substrates, but the pattern of the grooves can be varied from that of straight, parallel, equispaced grooves by changing the shape of the interfering wavefronts. In this way, circular, hyperbolic or elliptical grooves can be formed which impart to the grating different focal properties and these can be used to offset the aberrations of the optical system in which the grating is to be used. This leads to an improvement in the performance of the instrument and again offers the designer new scope.

☐   Replication

Until interference gratings became available in the early 1970s, practically all gratings sold for use in spectroscopic instruments were replicas. The cost of ruling a grating is so high and the time taken so long that it would be impracticable to use masters on a commercial scale. The invention of replication techniques may therefore be seen as one of the very important factors in the development of spectroscopy. Without such techniques instruments such as spectrophotometers which are made and used in large numbers would all have to rely on prisms as the dispersing element (some still do use prisms) and the higher-grade instruments which take full advantage of gratings would be available to the few who could afford them. Replication, therefore, has had a profound effect on the development of grating technology, not least because it offered the prospect of a better financial return on the large investment in a ruling engine.

Lord Rayleigh demonstrated that it was possible to replicate gratings by casting and such replicas had been incorporated into commercial instruments by Adam Hilger as early as 1910. However, it was not until the early 1950s that high-quality replica gratings became available. The basic process was patented by White and

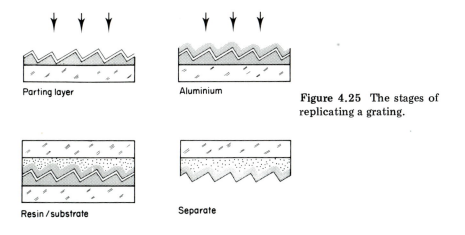

Parting layer

Aluminium

Resin / substrate

Separate

Figure 4.25 The stages of replicating a grating.

Fraser in 1949 and is shown diagrammatically in Figure 4.25. The master grating is first coated with a thin layer of some non-adherent material, the precise details of which are usually a closely guarded secret but gold, oils and silanes have all been used successfully (Torbin and Nizhin 1973). This is then followed by a substantial layer of aluminium. The optical quality of the surface of this layer is not important as it will be buried within the replica, but it is important to avoid pinholes which might allow resin to penetrate and cause irreparable local damage to the master when the replica is pulled off. A pool of resin is then poured onto the master, and the replica substrate, which may have been treated to improve adhesion, is placed on top. Again, the precise details are not usually published, but various epoxy and polyester resins may be used. When the resin has hardened the replica and master are separated. They may be forced apart with a wedge or they may be gripped by the edges and pulled apart. This process may be aided by thermal shock and if the coefficients of thermal expansion of the two blanks are different then a uniform change in temperature will also induce strain at the boundary and encourage separation.

The production of the first replica is the most critical state in the use of a master grating. Once this has successfully been achieved the replica may be used as a sub-master for the production of further generations of replica and the day-to-day production of gratings for use in instruments will use a sub-master which is several generations removed from the original. In this way the original is preserved and if a sub-master is either damaged accidentally or wears out it can be replaced. When large numbers of small gratings are required, it is usually more efficient to saw these up from a larger grating, in which case two generations are used. The sub-master and the final replica are sawn before separation so that each protects the surface of the other.

A replica will, of course, take up the converse shape of the master. A concave master will yield a convex replica but a plane master will be copied as a plane replica. Since the space between the replica blank is filled with resin it should, in principle, be possible to cast a replica on a blank of any shape. However, in practice, resins shrink perceptibly as they cure and this can lead to distortion in the shape of the replica.

Consider the case of a perfectly flat master which is to be replicated on a blank which departs from flatness by an amount x using a resin which shrinks by $y\%$ on curing. If the shrinkage is uniform the total thickness of the resin does not matter and the departure

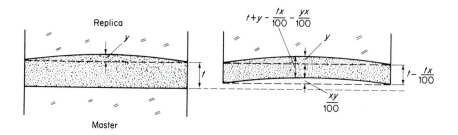

Figure 4.26 The effect on a replica of shrinkage in the resin.

from flatness of the replica is simply $xy/100$. If the shrinkage is non-uniform then this will cause distortion, even with a perfect replica blank, so it is advisable to keep the thickness of the resin to a minimum. For replicas of the best optical quality it is necessary to use substrates that are worked, albeit not to the same tolerances as the master. If the shrinkage of the resin is, let us say, 5%, then we require a replica blank which is flat to one wavelength in order to achieve a replica which is accurate to $\lambda/20$.

The inversion on replication of the features of a grating applies to the groove profile as well as to the overall shape and this can lead to differences between the spectral performance of odd and even numbered generations. For example, in the case of a ruled grating, it must be remembered that the shape of the apex of the groove is determined by the formation of the shallow and steep facets of adjacent grooves, whereas the bottom of the groove is determined more by the shape of the diamond. It frequently happens, therefore, that the bottom of the groove is smoother and better formed than the top and if the grating is used in such a way that the bottom of the groove is shadowed, clearly a better performance may be expected from an odd generation (where the defects are at the bottom) than from the master or an even generation of replica. This is especially true in the case of echelles, where either one has to take particular care in setting up the diamond in order to form the apex well, or one has to ensure that only odd-generation replicas are used in instruments. Similar considerations may also apply to the interference gratings where non-linearity in the resist may lead to groove profiles which are asymmetric with respect to the mean grating plane (Wilson, McPhedran and Waterworth 1973).

5

The Spectroscopic Properties
of Gratings

In Chapter 2 we described a grating as a Dirac comb multiplied by an aperture function and convoluted with a groove. The Dirac comb, which describes the position of the grooves, determines where the diffracted orders lie. The aperture function determines the degree of perfection of the focus of the spectrum and hence the extent to which the theoretical maximum resolving power is realized. The form of the groove determines the way in which light is distributed among the diffracted orders. These features determine the three practical aspects of a grating's performance which are of interest to a spectroscopist and which depend upon the quality of the grating. These are, respectively, the purity of spectrum, the resolving power and the efficiency, each of which we shall consider separately.

In order to specify the performance of a grating we need both to define and to measure its properties. For each parameter we can do this in two ways: first by studying the grating itself and second by studying the spectrum (i.e. the Fourier transform of the grating) directly. Both are equally valid provided that one has sufficient information and an adequate theory to convert from grating parameters to spectral characteristics. In general it is the spectral characteristics that interest the spectroscopist, but measurements of the grating itself are often most useful to the manufacturer in the diagnosis of faults. Owing to the inverse relationship between the dimensions of an object and those of its Fourier transform, or in our case between a grating and the spectrum, small-scale features of the spectrum correspond to large-scale features of the grating and vice versa, and it is usually easiest to measure the large-scale parameter. It is easier, for example, to measure the shape of the diffracted wavefront than to measure resolving power directly, and it is usually easier to measure the efficiency rather than the groove profile.

☐   Resolving power

In theory the resolving power of a grating is determined by its size and the wavelength at which it is used, or by the total number of grooves and the order in which it is used. In practice the resolution achieved in an instrument will depend upon the design of the instrument and upon the quality of the optical components, including the grating. It may well be limited by factors other than the grating, so when discussing the resolving power of the grating we shall assume that it is used in conjunction with a perfect optical system. Under these circumstances the fraction of the theoretical resolving power that the grating achieves is limited by aberrations in the wavefront; in this respect the grating is no different from any other optical component. For the sake of simplicity, and because concave gratings will be discussed in Chapter 7, we shall confine our attention to plane gratings. In our hypothetical perfect instrument the grating is illuminated with a perfectly flat wavefront. In order to produce a perfect image in the spectral plane, the grating is required to generate a perfectly flat diffracted wavefront.

It is, therefore, possible to specify and to measure this aspect of a grating's performance in two ways: either one can measure the flatness of the diffracted wavefront or one can measure the shape and width of a spectral image. The latter method has the appeal that it provides the spectroscopist with the information in the form in which he will use it, but it entails considerable practical difficulty. First of all it is necessary to test the grating in an instrument which does not introduce significant aberrations. Otherwise the instrument will itself distort the spectral image; the results will only be valid for that instrument and will be irrelevant if the grating is to be used in a different one.

Overcoming this problem requires careful design and construction of the test bench. In order to use slits of manageable widths when testing large gratings, it is usually necessary that the focal length of the system be rather long. It is also far easier to obtain diffraction-limited optical components such as concave mirrors with small numerical aperture (i.e. ratio of width to focal length) so if one wishes to test large gratings this again implies long focal lengths and a test bench that is very much larger than the average instrument. For example, the recently commissioned grating test bench at the National Physical Laboratory (Verrill 1981) has an aperture of 325 mm and an equivalent focal length of 10 m and the optical system has to be folded in order to accommodate it in one room. When working over this sort of distance ground vibrations, temperature gradients and draughts are particularly troublesome and great care must be taken to reduce or eliminate them.

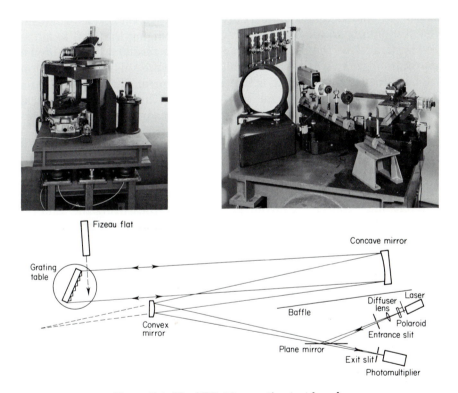

Figure 5.1 The NPL 10 m grating test bench.

Figure 5.1 shows a photograph and a diagram of the NPL test bench and illustrates the sort of equipment that is required to make adequate measurements of resolving power. The instrument must be set up with care, to ensure that the beam incident upon the grating is well collimated, to find the position of best focus and to ensure that the entrance and exist slits are parallel. If an optical flat is mounted in the instrument it is possible to find a good focus with imperfectly collimated light. However, when the grating is substituted it may give rise to chromatic errors because, since the grating is inclined to the beam, different positions on the grating correspond to different optical paths and define the different positions of best focus for the image. If the diffracted wavefront is curved, owing either to curvature of the blank or to ruling errors, then when the grating is substituted for the flat there will be a shift in the position of best focus. One then has the choice of measuring the instrumental linewidth either at the focus corresponding to a flat or at the best focus for the grating. It can be argued that spherical curvature or even vertical astigmatism (corresponding to a cylindrical wavefront bringing the light to a line focus parallel to the slit) will not impair the resolving power of the grating and that it is therefore valid to measure this at the position of best focus. However, it is not always convenient to refocus the instrument to accommodate this curvature. In some astronomical instruments, for example, the gratings are changed several times during the course of a night's observing and it would be too time-consuming to refocus the instrument each time. In this case the resolving power measured at the "flat focus" is most relevant.

Until stabilized lasers were available, one of the major problems in the direct measurement of resolving power was the search for a suitable light source. A favourite method of presenting the data on resolving power was to show a high-resolution spectrum of some emission line which contained a considerable amount of hyperfine structure and the quality of the grating was judged by the number of components that were resolved. Unfortunately, this was a very unsatisfactory basis for the comparison of gratings, especially when they were measured in different laboratories, because the results depended as much upon the light sources as they did upon the gratings. Now that frequency-stabilized helium–neon lasers are readily available the problem does not arise, because it is possible to perform the measurements with light which has a bandwidth much smaller than that which can be resolved by the grating. The measured linewidth is then determined effectively by the instrument and by the grating. It is interesting to note, incidentally, that the full width at half-height of the instrument function is in fact rather insensitive to defects in the grating and although it has been widely used to justify claims of excellence by grating manufacturers it is a somewhat unsatisfactory criterion. The reason for this is that many common defects tend to send light that should have gone into the central maximum into the wings of the line as shown in exaggerated form in Figure 2.31. In fact, the height of the central maximum would be a far more sensitive test, but this involves other variables and would be more difficult to measure in practice (Verrill 1980b).

A further test which is frequently used to check the quality of gratings is the Foucault knife-edge test. The grating is mounted in the test bench, a knife-edge is placed at the focus so that the main image is blocked off, and the grating is viewed from behind the knife-edge. For a perfect grating all of the light goes into the image, which is obstructed, so the field of view is uniformly dark. However, imperfections in the grating give rise to variations in the direction of propagation of the wavefront so that light passes by the knife-edge and the corresponding area of the grating appears bright. This technique does not readily give quantitative information, but with an experienced operator it can very quickly give reliable information for the purpose of quality control and the diagnosis of faults.

☐   Interferometric assessment of the diffracted wavefront

The most convenient means of investigating the resolving power of a grating is to measure the form of the diffracted wavefront on an interferometer. This compares the actual diffracted wavefront with an ideal or known wavefront and the difference between them is deduced from the interference fringe pattern that is generated. This is a test not so much of the resolving power but of the quality of the ruling and of the blank. However, given this information and Rayleigh's criterion that departures from the ideal wavefront of $\lambda/4$ or less are unlikely to cause a perceptible degeneration of the image, it is often sufficient to know the resolving power of the grating is "diffraction-limited". In a great many applications the grating is never used anywhere near the theoretical limit of resolution, so that a cursory study of the wavefront is often sufficient to reassure the user that the grating will not degrade the spectral image. In those applications where high resolution and the accurate measurement of wavelength is of importance, the interferograms must be studied in more detail.

An interferogram is in effect a contour map of the test wavefront in which the contour interval is equal to the wavelength of light. A dark band corresponding to destructive interference occurs whenever the path difference between the two wavefronts is equal to an odd number of half-wavelengths, as shown in Figure 5.2. When the match between the reference and the test wavefront is good, let us

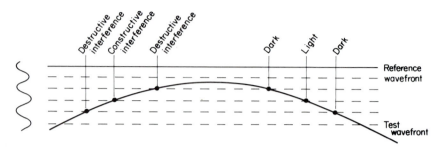

Figure 5.2 An interferogram as a contour map of a wavefront.

say half a wavelength or better, then the field of view will be of nearly uniform intensity and it will be difficult to extract the necessary information. It is therefore customary to introduce tilt between the wavefronts. The pattern then consists of a number of parallel straight fringes and differences between the shapes of the test and reference wavefronts will be seen as deviation from straightness of these fringes. The interferogram tells us the relative phase at any point on the aperture of the grating and is particularly useful in the diagnosis of faults because it pinpoints directly the areas of the grating which are in error. If, for example, a ruling engine momentarily lost servo control, then there would be a step in the diffracted wavefront. If the temperature control system failed to operate, there may well be diurnal variation in groove spacing and this will be seen as an error in the interferogram with a period equal to one day's ruling. If there is a particular area of the grating which has a poor wavefront, then this can be seen immediately from the interferogram, whereas a direct measurement of the resolving power would only indicate that the grating was faulty in some way. Figure 5.3 shows a sample of interferograms and their interpretation (Learner 1972).

It is possible to use the information contained in an interferogram to determine the profile of the spectral image and hence give us direct information about the resolving power. The phase relative to some arbitrary datum is measured at a large number of points on the diffracted wavefront and is then used to calculate the Fourier transform of the aperture function.[†] Such calculations are particularly important when the grating is to be used at the maximum possible resolution, as for example in the precise comparison of wavelengths. From interferograms taken at a single wavelength it is possible to

[†] Ideally such calculations should also take account of variations of efficiency across the grating.

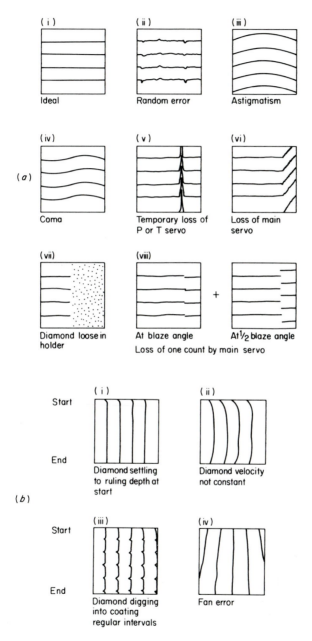

Figure 5.3 Grating wavefront interferograms and their interpretation (after Learner 1972). (a) Fringes across grooves ("horizontal"); (b) fringes along grooves ("vertical").

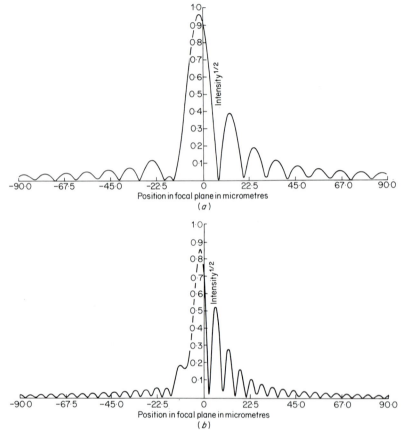

Figure 5.4 Distortion of the diffraction pattern at different wavelengths due to cyclic errors in the wavefront. (*a*) Diffraction pattern from a single-cycle sine wavefront of amplitude $\lambda/10$ at $\lambda 632.8$ nm, observed at $\lambda 632.8$ nm. (*b*) Diffraction pattern from a single-cycle sine wavefront of amplitude $\lambda/10$ at $\lambda 632.8$ nm, observed at $\lambda 316.4$ nm.

calculate the profile of spectral images formed at different wavelengths and it is interesting to note that even with a grating of high quality this profile is by no means constant, as is shown in Figure 5.4.

The danger is that, when comparing widely differing wavelengths, the change of profile will cause a shift in the "centre of gravity" of the spectral line and this must be taken into account if it is not to result in errors with measurement of wavelength (Preston 1970).

There are three types of interferometer that are commonly used in the testing of plane gratings. In each case it is necessary that the grating is tested in collimated light. For gratings of sizes that are in common use, Twyman–Green and Fizeau interferometers are available commercially and are convenient to use. For larger gratings it is often necessary to use either a Fizeau interferometer incorporated into a purpose-built grating test bench or a common-path interferometer which may be set up using the instrument for which the grating is intended.

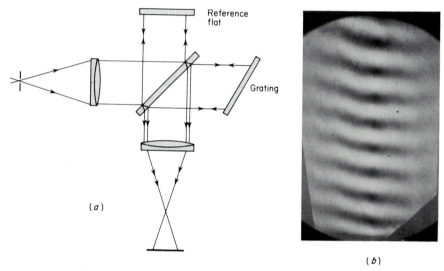

Figure 5.5 (a) The Twyman–Green interferometer; (b) an interferogram taken using an Hg lamp.

In the Twyman–Green interferometer as shown in Figure 5.5, a collimated beam of monochromatic light is split into two components by a semi-reflecting mirror. One component is reflected back down its own path by a good plane mirror and serves as the reference wave-front, while the other is returned along its own path by the grating under test. The two beams are combined at the beam splitter and the interference pattern may either be photographed or inspected visually. If, as is often the case, an incoherent light source is used, it is necessary to adjust the path lengths of the two beams to be equal in order to obtain the best fringe contrast. For small gratings a thermal source such as a low-pressure mercury lamp will have sufficient coherence to give a useful interferogram, but since the whole point of a grating is that the path length is different for each groove, the path lengths in the interferometer can only be equalized at one point on the grating width. Elsewhere, as we see from Figure 5.5(*b*) there is a loss of contrast. This problem can be overcome by using a suitable laser as the light source, but this often introduces additional problems due to interference of multiple reflections within the system.

Given adequate coherence, the fringe contrast depends upon the relative amplitudes of the two beams and provided that the grating is reasonably efficient adequate contrast is usually achieved with a fully reflecting reference mirror. Since each beam suffers one reflection and one transmission at the beam splitter, the contrast is not affected by the efficiency of the semi-reflecting coating. The disadvantage with the Twyman–Green interferometer is that it requires a beam splitter which is the same size as the grating under test and that this must be of very high optical quality. One beam makes two more passes through the glass than the other beam and any aberrations due to inhomogeneity of the beam splitter blank will be recorded on the interferogram.

One instrument which is commonly used to test the flatness of diffracted wavefronts is the Fizeau interferometer, shown diagrammatically in Figure 5.6. Here the functions of reference flat and beam splitter are performed by a single optical flat, which may be used uncoated or may have some partially reflecting coating applied to the reference surface. A collimated beam of light passes through the back of the reference flat and a proportion of this is reflected back by the reference surface. The rest continues and is diffracted or reflected back by the specimen under test. The two reflected wavefronts interfere and may be viewed near the focus or by means of a small thin beam splitter. The great advantage of this system is that both beams follow the same optical path so that any aberration due to the optics in one beam will be present in the other and its effects will therefore cancel. This is only strictly true if the gap between the reference flat and the specimen is negligible. If, due to aberrations or poor collimation, any portion of the beam propagates in a different direction from the rest, it will undergo a lateral shift by the time it has been reflected back from the specimen. It will therefore interfere not with its counterpart reflected from the reference flat but from that of an adjoining area and the effect of the aberrations will not cancel. The amount of shift will depend upon the size of the gap and with a grating this varies across the aperture and cannot be made small. It is, therefore, necessary when testing gratings to take rather more care to ensure good collimation of the incident beam than is usually necessary when testing flats.

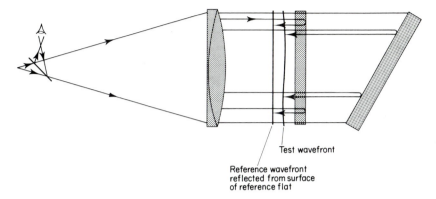

Test wavefront

Reference wavefront
reflected from surface
of reference flat

Figure 5.6 The Fizeau interferometer.

Nevertheless, the system remains a very practical one for most applications.

There are, however, disadvantages. The first is that it requires a test flat of reasonable optical quality in transmission to be of the same size as the grating. The reference flat on the NPL grating test bench is of 350 mm diameter and will accommodate most gratings, but few commercial interferometers have an aperture as large as this. The second problem concerns the reflectance of the reference surface. The contrast of the fringes generated between the 4% reflection of the reference beam and, let us say, 60% reflected by the grating is rather low. This can be improved by enhancing the reflectance of the reference flat, but then multiple reflections between the flat and the specimen become noticeable and can distort the fringe pattern (Clapham and Dew 1967). The optimum reflectance for the flat is therefore difficult to determine and in practice, rather than run the risk of altering the figure of the flat by uneven coating, it is often left uncoated. Despite these disadvantages, the Fizeau interferometer remains a widely used and convenient system for the testing of gratings.

A third system which is particularly useful when testing large gratings is the common-path interferometer shown in Figure 5.7 (Learner 1972). In this case a small incident beam is divided near the focus of the collimator. One part is expanded to fill the collimator, is diffracted by the grating and is brought again to a focus near the beam splitting prism but misses the expanding lens. The other part remains unexpanded and it follows the same optical path, but via a pair of small plane mirrors. It reaches the expanding lens on its return to the beam splitting prism where it is expanded to match

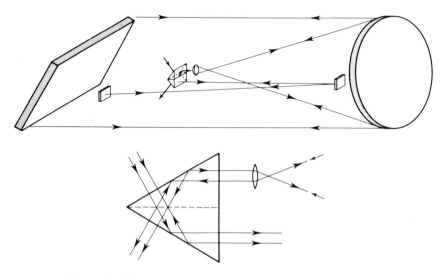

Figure 5.7 The common path exploded shear interferometer.

and interfere with the first beam. The advantage of this system is that it requires the smallest number of large optical components, since both the beam splitter and reference flats are smaller than the grating aperture. Furthermore, the amplitudes of the two beams may be equalized by the appropriate choice of reflectance of the mirrors in the reference beam. There is the disadvantage that the large collimating mirror must be of excellent quality, since aberrations from this component will not cancel as they do to some extent in the Twyman–Green and Fizeau interferometers. However, if a grating is to be used at maximum resolution, one can assume that a suitable mirror must be available as part of the instrument and this can be used to test the grating.

☐   Spectral purity: effects of diffraction

For an ideal grating of infinite extent one would expect light to be diffracted only into the directions determined by the grating equation. However, even for a perfect grating, Fraunhofer diffraction gives rise to $N - 2$ secondary maxima of intensity between the main orders. Any light which is observed between the main orders and is in excess of that accounted for by Fraunhofer diffraction is regarded as "spectral impurity" and arises from some defect of the grating or the instrument. For the moment we shall assume that the instrument is perfect and that the other components contribute nothing to the level of stray light. This assumption is by no means always valid, but at present we are concerned just with the gratings. There are several distinct possible sources of stray light and each gives rise to a different form of unwanted scatter in the spectrum. These are summarized in Figure 5.8, in which we assume that a grating is illuminated by a collimated beam from a point source of monochromatic light and that this is then focused to a series of points in the spectral plane. In this context it is relevant to consider the effects of the way in which even a perfect grating is illuminated. The spectroscopist wishes to have as little light as possible between the diffracted orders and from his point of view it does not matter whether the light comes from an imperfection of the grating or from diffraction effects.

The intensity distribution in the plane of the spectral image for a perfect, finite grating with a uniformly illuminated square aperture was shown in Chapter 2 to be given by the function:

$$I_\theta = a^2 \frac{\sin^2 N\gamma}{\sin^2\gamma} \quad \text{where} \quad \gamma = \frac{\pi d \sin \theta}{\lambda}$$

If the resolution of the optical system is inadequate to resolve the individual secondary maxima, the mean measured intensity will be equal to half that at the peak of the maximum and will be given by[†]

$$\frac{I_0}{2}\left[\frac{2}{\pi(2p+1)}\right]^2 \simeq \frac{I_0}{20p^2}$$

where p refers to the pth secondary maximum.

If the grating is fully and uniformly illuminated, this level of apparent stray light is the minimum that will be observed even with a perfect grating, and if the grating is not the limiting aperture in the system, then Fraunhofer diffraction of any other limiting aperture will give rise to a similar amount of light in the extended wings of the diffracted maxima.

If, as in the second case in Figure 5.8, the aperture is circular, then a similar amount of energy goes into the wings of the diffraction pattern, which in this case is similar to an Airy disc (for a circular grating the effective aperture will be elliptical because the grating is inclined to the beam). However, this energy is no longer concentrated along the line joining the orders, but is spread over a far greater area. Its intensity (i.e. flux per unit area) is therefore

[†] For a maximum: $\quad \gamma = \dfrac{p\pi}{N} + \dfrac{\pi}{2} \quad I_0 = N^2 a^2.$

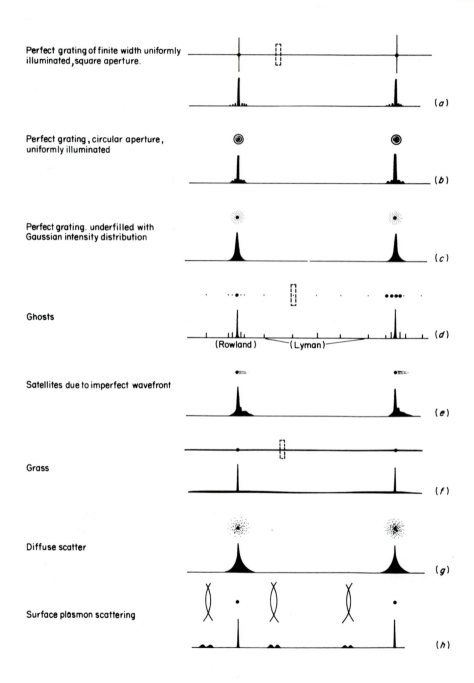

Perfect grating of finite width uniformly illuminated, square aperture.

(a)

Perfect grating, circular aperture, uniformly illuminated

(b)

Perfect grating. underfilled with Gaussian intensity distribution

(c)

Ghosts

(Rowland) (Lyman)

(d)

Satellites due to imperfect wavefront

(e)

Grass

(f)

Diffuse scatter

(g)

Surface plasmon scattering

(h)

Figure 5.8 Forms of spectral impurity.

very significantly reduced and the amount of flux passing through an exit slit is similarly diminished except when the slit is very close to the diffracted order. The level of diffracted light can be reduced still further by illuminating the grating not uniformly but with a gaussian distribution of intensity. Here again, the diffracted light is distributed over a far wider area although as the effective width of the grating is reduced there is a corresponding increase in the spectral linewidth.

Most gratings are square or rectangular partly for historical reasons, because that is the format produced by a ruling engine, and partly because there is insufficient space available in many instrumental designs to allow for circular blanks. Furthermore, it is most usual for the grating to be fully illuminated. This is partly because with ruled gratings Fraunhofer diffraction was seldom the limiting factor (as it was overshadowed by grass) and partly on economic grounds. To apodize, that is to control in this way the distribution of light in the aperture, involves cutting off some of the useful area of the grating with a consequent loss of light-gathering power. Since the grating is usually the most expensive component in the optical system, it is customary to use it as fully as possible. However, even if the reduction of Fraunhofer diffraction by apodization is not feasible in many practical instruments, it is a most valuable technique when testing gratings because it enables the effects of diffraction to be separated from the effects caused by defects in the grating and thus allows a better assessment of the quality of the grating itself (Jaquinot 1964).

□   Spectral purity: grating defects

With ruled gratings spurious images are frequently observed in the spectral plane. These are caused by periodic variations in the position or shape of the grating grooves. Such images are called *ghosts* and two types are normally distinguished. "Rowland ghosts" lie close to the main diffracted orders and are caused by comparatively long-term errors corresponding to errors in the lead screw or gears in the ruling engine. "Lyman ghosts" on the other hand, are found further away from the parent line and are caused by short-term errors with a period of a few grooves (Lyman 1901, Wood 1924, Gale 1937).

Optically there is no difference between Rowland ghosts and Lyman ghosts. The only difference lies in their period and in their origin. Both are due to periodic variations of groove depth or position, so that in effect the grating aperture function is multiplied by a periodic phase and amplitude function which behaves as a grating in its own right, and the spectrum of the "error" grating is then convoluted with the spectrum of the main grating. In practice Lyman ghosts are often more difficult to detect and because they are so far away from the main orders they are often confused with details of the spectrum.

The measurement of the intensities of ghosts is fairly straightforward. Since the ghost is a genuine spectral image it has the same shape as the main diffracted order and its intensity relative to that order can be measured directly. It matters not whether total flux or peak intensity is measured, nor does the measurement depend upon the resolution of the instrument. Provided that the ghost and the parent order are measured under the same conditions the comparison should be valid. If there are errors in the diffracted wavefront that are not periodic but occur over a substantial fraction of the grating aperture, then this results in a distortion of the Fraunhofer diffraction pattern of the aperture. Apart from the distortion of the central maximum, which would give rise to a loss of resolving power, the secondary maxima are also distorted in both position and magnitude. This gives rise to defects known as *satellites*, an example of which is shown in Figure 5.9. To some extent they resemble Rowland ghosts in that they are close to the exciting line, but they differ in that they are not necessarily disposed symmetrically about the main order. They are particularly troublesome in high-resolution absorption spectroscopy because they tend to "fill in" an absorption line and lead to errors in its measured "intensity".

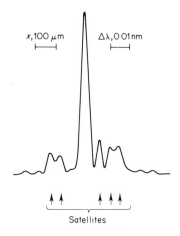

Figure 5.9 Satellites.

Shorter-term random errors in the wavefront give rise (in accordance with the reciprocity between the grating and image planes) to scattered light further away from the main order. If these errors occur in a direction perpendicular to the grooves, that is they arise from variations of the position of identical grooves, then this light will be distributed in a direction perpendicular to the grooves and will be seen as a faint line joining the diffracted orders (shown in Figure 5.8(f). This is referred to as *grass*. The random distribution of errors can be considered equivalent to a Fourier sum of periodic functions of suitable period, amplitude and phase. Grass can be considered as the sum of the ghosts which each of these periodic functions generates.

If the random errors have no preferred direction, then the scattered light which they generate will be distributed into all directions. This pattern of diffuse scattering will be convoluted with the pattern of the diffracted orders as shown in Figure 5.8(g). Scattering of this type is caused by roughness on the surface of the grating, and by dust, scratches or any other random imperfection and is the equivalent of veiling glare in lens systems. For interference gratings such random errors can, unless great care is taken during manufacture, arise from spurious fringes caused by interference from light scattered from the components of the interferometer. In effect, the grating is a hologram of the apparatus and the reconstruction of this can be a source of stray light in the spectral plane. In practice it is only in the most demanding of applications that such effects are troublesome.

A final form of stray light is shown in Figure 5.8(h). This is called surface plasmon scattering and is due to a complex interaction between the electromagnetic field of the incident light, the conduction electrons of the metal and the roughness of the grating surface. It is a resonant phenomenon which will be discussed in more detail in Chapter 6. Its description is beyond the capability of the simple Fourier transform approach that we have applied to other forms of stray light and, unlike the other features, it is dependent upon polarization and upon the material of the grating surface. For the moment we merely note its existence.

It is evident from what has been said so far that the light detected at a point in the focal plane of an instrument between the diffracted orders will probably have arisen from a variety of sources. Just how much "stray" light is detected will depend upon the relative importance of the various sources and how the measurement is made. Let us consider the case depicted in Figure 5.8 in which one point on the entrance slit is illuminated with monochromatic light. The first question we must ask is how the "spectrum" and the stray light are to be detected. In a spectrograph the light is recorded on a photographic plate or its equivalent in which the response is determined by the irradiance of the light in the focal plane, that is the *flux per unit area*. In a monochromator the light is detected photoelectrically and the response is determined by the *total flux* passing through the exit slit and falling onto the sensitive area of the detector. It therefore follows that those forms of stray light which are concentrated into a comparately small region will have a proportionately greater effect in a photographic measurement than they will in a photoelectric measurement made with large slits. On the other hand, if the entrance slit is fully illuminated (as is usually the case with conventional sources) then the distinction between concentrated and diffuse scattering is less marked, because the diffuse scattering from neighbouring regions will contribute to the flux detected at any point on the exit slits.

☐   The measurement of stray light

The photoelectric measurement of stray light is a fairly straightforward matter. The instrument is illuminated with monochromatic light from a laser and the grating is rotated as if to scan the spectrum. The flux passing through the exit slit at any given wavelength setting is measured with a photomultiplier and compared with the flux in the main diffracted order, measured when the wavelength setting corresponds to that of the incident light. What is not straightforward is the interpretation of the data obtained in this way. There is no problem if all that is required is a knowledge of the performance of a given grating in a given instrument under a given set of conditions, but difficulties arise when it is required to assess the performance of a grating in order to predict its performance in another instrument or to compare it with a grating of a different type. The amount of stray light that is measured depends upon many factors, all of which must be specified and taken into account when interpreting the data. First of all the Fraunhofer diffraction depends upon the size of the grating and the way in which it is illuminated. For a small uniformly illuminated rectangular grating there may be regions where diffraction predominates over the stray light due to imperfections in the grating. The amount of flux detected at the exit slit will depend upon the quality of the grating (which is ultimately what one hopes to assess) and the size and shape of the exit slit.

Ghost intensities may be specified quite easily, because a ghost image is the same shape as that of the main diffracted order and a simple comparison of irradiance or flux is valid. Grass, on the other hand, extends more or less continuously from one diffracted order to the next. The proportion of flux from this source relative to that in the main order will increase linearly with the width of the exit slit provided that this is wide enough to accept the whole of the image of the main order, and the measurement is independent of the height of the slit. This contrasts with the diffuse forms of scattering, in which the amount of light detected is proportional to both the width and the height of the slit. It follows that, whenever measurements of stray light are presented, it is necessary to specify both the nature of illumination (e.g. underfilled or overfilled) and the angular subtense of the slit, but unless the form and spatial distribution of the stray light is also known, there will not be sufficient data to make a general comparison of two gratings. Let us suppose, for example, that one wished to compare a ruled grating and an interference grating of the same size, shape and pitch. Most of the scattering from the ruled grating will be grass, whereas that from the interference grating will be diffuse. We will assume that the exit slit has the same width in both cases. A short slit will pass most of the grass from the ruled grating but accept very little diffuse light from the interference grating and the comparison will give the most favourable result for the interference grating. As the exit slit height is increased, the measured amount of diffusely scattered light increases, but the level of grass stays constant. So the comparison becomes less favourable to the interference grating and in extreme cases, if the exit slit is very long, the diffuse scattering from the interference grating can exceed the grass from a ruled grating.

This teaches us two things; firstly that it is an advantage to use slits that are as short as possible consistent with the need for the instrument to accept the maximum flux. Secondly, any comparison between gratings should be treated with great caution unless the measurements are all made in the instrument for which the grating is intended. Unfortunately, there is no generally accepted convention by which different manufacturers can measure and specify the stray light performance of their gratings, even though they are often pleased to quote ghost intensities, resolving power and efficiency.

The problem becomes even more complicated when one seeks to compare gratings of different pitch. For example, it may be that the best quality ruled gratings for use in the visible might have 1200 grooves mm^{-1}, whereas in order to optimize efficiency one might choose an interference grating with 1800 grooves mm^{-1}. One now has to consider how wide the slits should be in order to make a fair comparison. Should they be the same physical width or should they be equal in terms of spectral bandwidth, in which case they would be approximately $1\frac{1}{2}$ times wider when measuring the interference grating. In this particular case one can choose between two ways of using the extra dispersion of the 1800 grooves mm^{-1} grating. Either one can increase the slit width and hence increase the amount of light entering the instrument without sacrificing resolution, or one can keep the slits the same width and reduce the proportion of stray light that one accepts. Whichever choice is made, it is important to bear in mind that the relative merits of gratings of different pitch will depend upon whether the measurements are made under conditions of equal luminosity or equal effective resolving power (Verrill 1979).

Unfortunately there are still further complications. Although it is a straightforward matter to measure the amount of light at different wavelength settings when the grating is illuminated with, say, a helium–neon laser, the results cannot easily be related to scattering at different wavelengths of illumination. Depending upon how the measurement is made, scattering from different sources may vary with wavelength in different ways, so a comparison of two different types of grating made with a helium–neon laser may well not be valid for a grating that is to be used in the ultraviolet (Sharp and Irish 1978).

It is evident that it is very difficult to provide a quantitative specification of stray light in a way that would enable the performance of a grating in one instrument to be predicted from measurements made in a different instrument. This is not to say that measurements and comparisons cannot be made, merely that conclusions drawn from them should be treated with great caution. Measurements of stray light can, in principle, be made in any form of spectrometer; the simplest method is to measure the performance of the instrument using the grating one wishes to test. However, if we wish to measure the performance of the grating itself, rather than that of the instrument as a whole, it is necessary to take particular care to ensure that the stray light that is detected originates from the grating and not from other parts of the instrument. One of the simplest tests of this is to look into the instrument through the fully opened exit slit either with the unaided eye or with a small telescope, taking care to ensure that the main diffracted order of the laser input is not passing through the exit slit at the time. In this way one views the instrument as if it were illuminated by stray light. Only those components which generate stray light will be visible and it is possible to make an immediate assessment of the relative importance of the various sources.

This is illustrated in Figure 5.10, which shows an interference grating mounted in a 3.5 m Littrow test bench. From this it is evident that of the total scattered light detected, a very significant proportion is generated from the collimating mirror. In fact the Littrow mounting as shown is a poor design from the point of view of stray light, because backscattered light is brought to focus near the exit slit. A far better arrangement is the Czerny–Turner spectrometer in which the entrance and exit slits are well separated.

 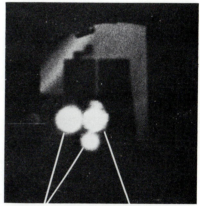

Scattering from collimator Scattering from grating

Room lights on   *Room lights off*

Figure 5.10 The grating and test bench viewed in scattered light.

Once it is established that the stray light is indeed coming from the grating, it is very instructive to focus a small telescope on the grating itself. In this way it is possible to pick out the individual areas which are giving rise to the stray light. The Fraunhofer diffraction pattern appears as a bright line along one or both edges of the grating. Figure 5.11 shows a selection of photographs of gratings taken in this way, together with a photograph of the spectral image and a photoelectric measurement.

Some of the gratings used as illustrations in Figure 5.11 were chosen because they exhibited the desired effect and not because

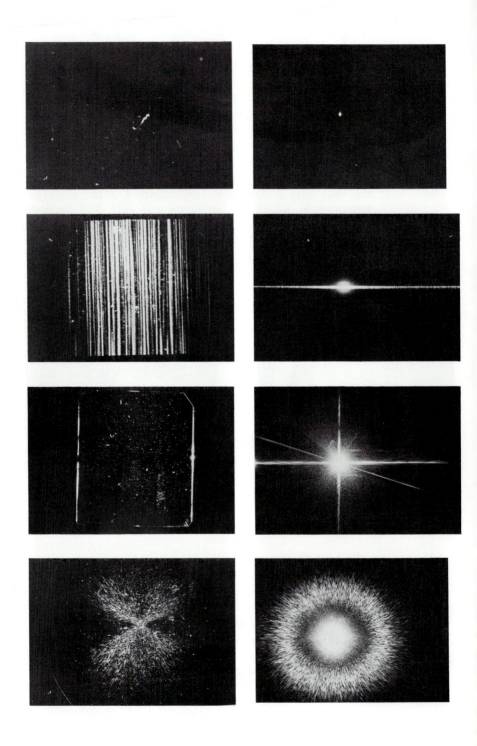

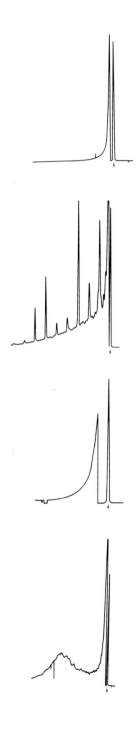

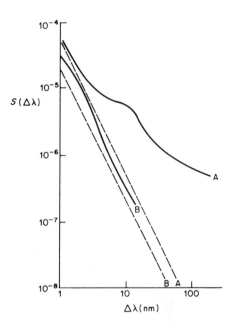

Figure 5.11 Stray light from various gratings. Colum A: viewed in the spectral plane. Column B: the grating viewed in the stray light. Column C: a linear photoelectric plot at 633 nm. Row (i): a good interference grating, underfilled. Row (ii): ghosts from a ruled grating. Row (iii): Fraunhofer diffraction from an overfilled square. Row (iv): interference grating on a badly coated photoresist surface exhibiting streaks.

Note: In figures A(i) to C(iv) the scales have been chosen to demonstrate particular features and are not comparable.

The inset shows a logarithmic plot of the stray light (solid lines) and the theoretical level of Fraunhofer diffraction (dotted lines) for two gratings measured at 633 nm under comparable conditions. (A) is a 1200 grooves mm^{-1} ruled grating 60 mm wide; (B) is an 1800 grooves mm^{-1} interference grating 80 mm wide. (After Verrill 1978.)

they were necessarily good gratings. However, the interference grating (*b*) and the ruled grating (*c*) were both chosen as good examples of their type.[†] The photoelectric measurements were adjusted to conditions of equal resolution (0.1 nm of spectrum) and where the slit height was one per cent of the focal length of the spectrometer (a realistic length). While bearing in mind the reservations we have discussed concerning the comparison of different types of grating, we may nevertheless regard these results as fairly typical of interference and ruled gratings. We note from these curves that the interference grating is significantly better than the ruled grating and that the improvement is greater the further one is from the main order. It is also interesting to note that beyond about 10 nm from the peak, the interference grating is virtually indistinguishable from a good plane mirror and that close to the peak, the contribution from Fraunhofer diffraction is much the same as that from scattering from the grating.

All that has been said so far has referred to the case in which the grating is illuminated with monochromatic light and the figures that have been quoted would be relevant to the study of fairly simple emission spectra consisting of a series of isolated lines. As the number of spectral lines increases so the contributions to the scattering will add and the level of stray light which is detected at any given point in the spectral plane will increase. In the limit one reaches the case in which the grating is illuminated with a broad-band continuum, for example of white light. Gratings are very frequently used under these conditions and stray light can be very troublesome, particularly in the case of quantitative absorption spectroscopy. Here, stray light has the effect of filling in strong absorption features. When the instrument is set on the wavelength corresponding to a strong absorption, the light that is transmitted consists of some light of the nominal wavelength and a complete broad-band spectrum of stray light. It is often important, particularly in astronomical spectroscopy, to measure the strength of an absorption band. This one does by measuring the proportion of light transmitted at the wavelength of the absorption peak. Unfortunately, this light cannot be distinguished from the continuum of stray light and serious errors in the measurement are possible.

Consider, for example, the case of the ruled grating in Figure 5.11. Let us suppose that the average level of stray light collected over an instrumental bandwidth of 0.1 nm is 10^{-6} at a given wavelength of

[†] In (*c*), its level of grass was low but ghosts were stronger than usual.

illumination. Let us further suppose that the grating is illuminated with a white light spectrum extending from 400 nm to 700 nm, i.e. 3000 times the bandwidth. The level of stray light measured at the same resolution would be 3000 times greater than that for monochromatic illumination, or about 0.3%. Thus an instrument working under these conditions would not measure any transmittance less than 0.3% (that is, any absorbance in excess of 2.5). This crude calculation does nothing more than suggest the order of magnitude of the problem and indicates the importance of correlating measurements made with monochromatic light with those made with white light. A complete correlation of these two types of measurement is a very complex matter. It is necessary first to know the way in which the level of stray light varies with the spectral distance from the main order for monochromatic light and to know how this varies with the wavelength of illumination. It is also necessary to know the spectral distribution of the source and the spectral response of the detector, so that an appropriate weighting factor may be assigned to all detectable wavelengths. Armed with this information it should then, in principle, be possible to calculate the amount of stray light at any given wavelength setting on the instrument.

It is possible to measure the stray light in an instrument by studying the way in which it tends to fill in and reduce the apparent strength of absorption bands. The instrument is illuminated with white light from which a single narrow band has been completely removed by a filter, as shown in Figure 5.12. Any light detected when the instrument is set at the absorption peak is due to stray light and can be measured directly. In effect this is equivalent to integrating the area under the stray light curve and gives a value which to a first approximation represents the total scattering in the whole of the spectral plane due to a single wavelength. This type of measurement is a convenient way of comparing one grating with another in a given instrument, but it is very difficult to extract quantitative information in a way which will relate to the performance of a grating when used in different instruments.

The integrated value of stray light will naturally depend upon the limits of integration, in this case the extent of the continuum with which the instrument is illuminated, and also upon the spectral response of the detection system. However, further useful information can be obtained by the use of band-pass or cut-off filters. The depth of an absorption band is first measured with no filter present and then again with a filter which cuts out all radiation below, say, 250 nm. The difference between the two measurements

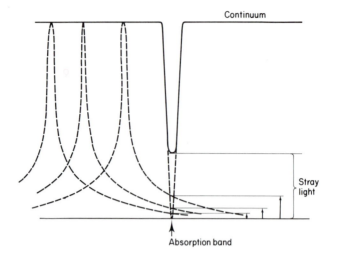

Figure 5.12 The effect of stray light on an absorption band.

corresponds to the level of scattering due to wavelengths below 250 nm. If the experiment is repeated with another filter which cuts off at 350 nm, the difference between the second and third measurements corresponds to the scattering of light between 250 nm and 350 nm. In this way it is possible to build up a useful picture of the influence of stray light in the instrument.

In practice, rather than use a highly absorbing filter, it is often convenient to use a liquid or gas absorption cell and to study the apparent deviation from Beer's law as the concentration is increased. Beer's law states that the intensity of light transmitted through a uniform distribution of material is given by

$$I_t = I_0 e^{-cla}$$

where l is the length of the cell, c is the concentration of absorbing material and a is the absorption coefficient. It follows that absorbance $(-\log_{10}(I_t/I_0))$ is proportional to concentration, as shown in Figure 5.13. However, if stray light is present, the measured absorbance will be less than expected as represented by the dotted line in the figure. The absorbance level at which the two curves diverge is a useful measure of the stray light in an instrument.

The value of the integrated level of stray light from a single wavelength will depend very strongly upon the lower limit of integration; that is, how close to the main line one starts. It is evident from

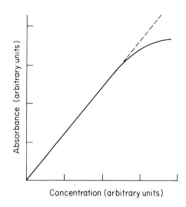

Figure 5.13 The effect of stray light on the measured relation between absorbance and concentration.

Figure 5.11 that the level of scattering is much higher close to the main diffracted order and this will have a strong influence on the measurement. If this region is to be included in an absorption experiment, the absorption band must be very narrow. In this case it is most important to ensure that the instrument is fully capable of resolving the band, because if it is not it will yield a pessimistic value of the stray light. In effect, in addition to filling-in from the bottom, the band will be filled in from the sides due to the lack of resolution (Figure 5.14).

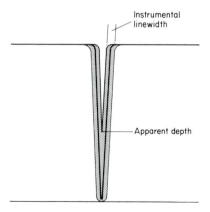

Figure 5.14 The effect of an instrumental linewidth which is greater than that of an absorption band.

☐   Efficiency

Probably the most sought after feature of a grating is its efficiency. It may be that in reality the ultimate limit of signal-to-noise ratio in an experiment is determined by the level of stray light, but a grating is unlikely to be widely used if it is not reasonably efficient. The efficiency of a grating can be defined in several ways and it is important to bear in mind the distinction between them. First the absolute efficiency is the percentage of incident light that is diffracted into the required order. This is the figure which is of most value to to the spectroscopist and unless otherwise stated this is what we shall mean whenever we refer simply to the "efficiency". It is determined by two factors, the shape of the groove and the reflectance of the material of which it is made. One encounters a second definition, which is the amount of energy diffracted into the required order compared, not with the incident energy, but with the amount of energy that would have been reflected by a plane mirror of the same material. This is known as the *relative* or *groove* efficiency and is a measure of the efficacy of the groove profile. Since the reflectance of the grating material is less than unity, it follows that relative efficiencies are always higher than absolute efficiencies. This should be borne in mind when considering specifications and data from grating manufacturers. Confusion between the two often arises in the visible region of the spectrum where gratings are usually coated with either bare or protected aluminium. This may have a reflectance of about 90%, in which case relative efficiencies are about 10% more optimistic than absolute efficiencies. Finally, in the third definition, the flux in the diffracted beam is compared to the total flux in the diffracted orders. This is more optimistic still since it neglects the loss of any scattered light. It is also rather misleading because, as we shall see, under some circumstances it can happen that significant fractions of the incident light are absorbed by the grating. In view of this we shall refer only to the absolute efficiency and the relative efficiency. The first helps the spectroscopist to calculate the throughput of his instrument and the second tells the manufacturer how successful he has been in generating the correct groove profile.

The complete knowledge of the properties of a grating calls for the measurement both of the groove profile and of its efficiency. The measurement of efficiency is a comparatively straightforward matter particularly for visible wavelengths. The measurement of groove profile, on the other hand, demands great care and attention, particularly if quantitative measurements are required. The user usually only needs to know the efficiency, but a detailed knowledge of the profile is very valuable to the manufacturer if he is ever to achieve high efficiencies by methods other than trial and error. This information is also necessary when it comes to making a detailed comparison between the measured values and the theory of grating efficiencies. As we shall see in the following chapter, it is only recently that a satisfactory agreement has been achieved between theory and experiment and the validation of the new theoretical results has required both a good knowledge of the groove profile and detailed measurements of efficiency.

☐ The measurement of groove profile

For comparatively coarse gratings used in the infrared it is possible to learn much from visible microscopy, especially interference microscopy. The widths of the grooves are many times the wavelength of visible light and it is therefore possible to resolve sufficient detail with a high-power microscope. The interference microscope is in effect a miniaturized version of the Twyman–Green interferometer and provides a contour map of the grating surface with a contour interval of half the wavelength. Figure 5.15 shows an interference micrograph of a coarse grating and indicates the sort of information one might expect to obtain.

For visible and ultraviolet gratings, where the groove spacing is often less than $1\,\mu$m, there is not sufficient spatial resolution for an optical microscope, and for this reason the electron microscope is widely used. The transmission electron microscope has the best resolution and can reveal detail as small as 0.5 nm. On the other hand, the scanning electron microscope has some advantages, even though it has an inferior resolution of the order of 20 nm. In practically all measurements using electron microscopy one cannot study the grating itself but has to take from it a small replica. This is somewhat time consuming and can introduce uncertainties in the results. Although the spatial resolution of the electron microscope is much higher than that of the optical microscope, it is often very difficult to obtain adequate information out of the plane of the object. Figure 5.16 shows a transmission electromicrograph of a ruled grating and, while this provides a great deal of detailed information about the structure of the surface of the grooves, it is very difficult to gain any impression of the shape of the groove. Various techniques

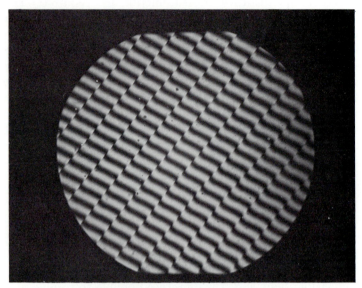

Figure 5.15 Interference micrograph of a coarse blazed grating.

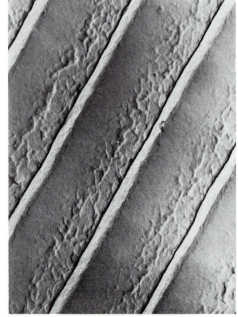

Figure 5.16 Transmission electron micrograph of a ruled grating (1200 grooves mm^{-1}).

have been used to overcome this. In the first a number of very small asbestos fibres are scattered on the grating which is then vacuum-coated with carbon at an oblique angle (Anderson *et al.*, 1965). The fibres then cast shadows which may be observed in the microscope; and from a knowledge of the conditions of shadowing, the shape of the groove can be calculated. There is likely to be some uncertainty in quantitative measurements if the fibres do not lie completely flat on the surface because of local irregularities. Nevertheless, the technique does give a good qualitative picture of the shape of the groove and Figure 5.17 shows a typical result.

Figure 5.17 Electron micrograph showing the shadow cast by an asbestos fibre. (Courtesy of E. G. Loewen, Bausch and Lomb).

Ideally we would like to view along the grooves so that the profile would be displayed directly. This can be achieved by bending a thin foil replica perpendicular to the grooves and viewing the profile as a silhouette. The technique can be used with both the transmission and the scanning electron microscope, but suffers from the disadvantage that by bending the replica one is liable to distort the form of the grooves. It is known from the spectroscopic performance of full-sized

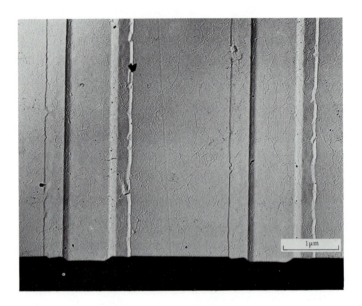

Figure 5.18 The superposition of a transmission electron micrograph and a silhouette of a shallow X-ray grating.

replica gratings that replication techniques can be remarkably faithful; but if the replica has to be distorted deliberately, then there is bound to be some doubt about how typical the observed grooves are. Figure 5.18 shows a standard transmission electron micrograph of a shallow X-ray grating with the silhouette superimposed upon it (Bennett 1969). A more direct method of viewing the edge of the grooves is to take a small replica on thin glass, to snap it across the grooves and to mount it in a scanning electron microscope with the grooves practically parallel to the electron beam (Brandes and Curran 1969). This can give results which are quite satisfactory for a qualitative assessment of the groove profile but it is very difficult to derive accurate quantitative information (Figure 5.19). The scanning electron microscope has a much lower resolution than the transmission electron microscope. This is degraded further because the sharp edges of the replica tend to charge up when illuminated by the electron beam, and this gives rise to a blur on the micrograph. The technique also calls for a certain care in aligning the sample and in interpreting the results. As can be seen from Figure 5.20, it is possible to observe an apparent blazed profile even with a symmetrical grating.

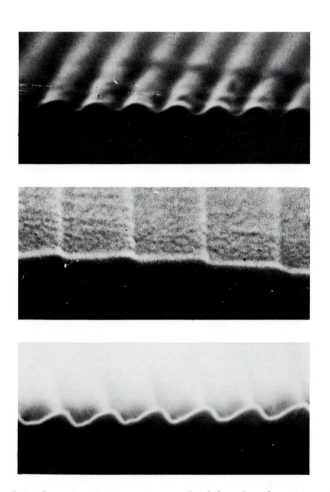

Figure 5.19 Scanning electron micrograph of the edge of grating replicas.

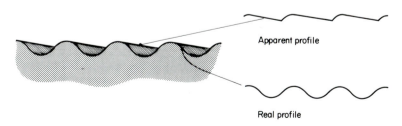

Apparent profile

Real profile

Figure 5.20 The possibility of misinterpretation of scanning electron micrographs.

An entirely different approach to the measurement of groove profiles involves the use of a "Talystep" step height measuring machine (Verrill 1973, 1975). This draws a lightly loaded diamond stylus across the surface of the grating and electronically monitors its vertical displacement as the tip moves up and down the grooves. The vertical resolution of this instrument is remarkably high: it can measure a step height of only a few Å, but its lateral resolution is limited by the size of the diamond tip. The stylus that is most commonly used in studying gratings is chisel-shaped, about $1.0 \mu m$ long and $0.1 \mu m$ or less wide with an included angle of $90°$. The width of the stylus sets a lower limit to the detail that can be observed because the trace that is generated is a convolution of the groove profile with the profile of the stylus tip. This also limits the maximum slope that can be traced, since if the angle of a groove face is greater than the angle of the facet of the stylus then the tip of the stylus loses contact with the surface of the grating, as illustrated in Figure 5.21(a), and the tip never reaches the bottom of the groove. In order to measure groove facet angles greater than $45°$ a special stylus may be used which has an included angle of $60°$ and is mounted on its shank at an angle of $20°$ so that the facets are inclined at $40°$ and $80°$. In this case particular care must be taken in the alignment of the chisel parallel to the grooves. Figure 5.21(b) shows a trace taken with such a stylus and the maximum angle here is $76°$. The main advantage of the Talystep is that it is very fast. A trace can be obtained in literally a few minutes, which is considerably less time than is needed to take a replica and mount it in an electron microscope. Because of this it is particularly valuable during the manufacture of gratings. As we saw in Chapter 4, studying the shape of trial rulings enables the loading and orientation of the diamond tool to be adjusted before starting to rule a grating. With interference gratings it is possible to develop and measure one grating before making the next exposure. The Talystep may be used with some confidence for gratings with pitches up to 1200 grooves mm^{-1}, although the results must be interpreted with care because, owing to the size of the stylus, the trace is not always a faithful representation of the profile. It can be used for even finer pitches provided that the grooves are shallow, but here one must be even more wary about the interpretation of the trace.

There is in effect no method which is suitable for measuring all features of all grating profiles and the problem generally becomes more difficult the finer is the pitch of the grating. Optical microscopy is useful, particularly for coarser gratings; transmission electron

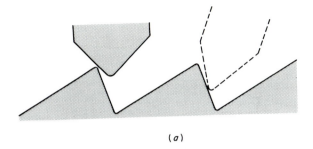

(a)

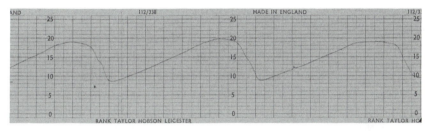

Figure 5.21 (a) The use of a Talystep to trace the groove profile; (b) groove profile measured with a tilted stylus.

microscopy gives excellent information concerning the structure of the surface of the grooves, but does not readily give information out of the plane of the grating. The Talystep, on the other hand, gives excellent vertical resolution, but its horizontal resolution is limited. The scanning electron microscope provides a general picture of the shape of the groove but does not lend itself to precise quantitative measurements. In order to study a groove profile in great detail it is advisable to use all methods. In practice, however, this effort is seldom justified since it is not so much the groove profile but the efficiency which is of most importance and, at least for visible and near UV radiation, this is easier to measure.

□   The visual estimation of grating pitch

In many cases it is possible for a person with normal colour vision to estimate with surprising accuracy the pitch of a grating simply by visual observation of the spectrum.

Figure 5.22 The visual estimation of grating pitch.

The grating is viewed in approximately a Littrow configuration with a light source directly behind the observer. The technique is first to view the source in the zero order and then to rotate the grating until the incident and diffracted rays just graze the surface of the grating. The number of spectra that are observed is noted and the wavelength corresponding to grazing incidence and grazing diffraction is estimated by its colour. In this configuration the dispersion tends to infinity, so a mere visual estimation is fairly accurate provided that there is not too much confusion of spectra due to overlapping orders.

Let us say that three spectra have been observed and that the third spectrum cuts off at a greenish yellow estimate to be about 550 nm. The spacing of the grooves will be three half-wavelengths, since in the Littrow case,

$$2d \sin \theta = m\lambda$$

so that $d = m\lambda/2$ when θ is 90°.

In this example the pitch is estimated to be $\frac{3}{2} \times 550$ nm $= 825$ nm or 1210 grooves mm^{-1}. So it is probably fair to assume that the grating had 1200 grooves mm^{-1}, which is a very common pitch. Another common pitch for gratings ruled on an engine controlled by a helium–neon laser interferometer is 1264 grooves mm^{-1}. In this case the cut-off wavelength would be about 526 nm which is green and should easily be distinguished from the greenish yellow of 550 nm by an observer with normal vision. This method works reasonably well for gratings of this sort of pitch and finer but is difficult to apply to gratings much coarser than 600 grooves mm^{-1} because the overlapping of the various orders gives rise to colours that are not pure spectral colours. Here, the best one could do is to use a monochromatic source (or one which had a few recognizable emission lines, such as the mercury lines in a fluorescent striplight) and count the number of orders.

☐   The measurement of efficiency

Measurements of efficiency are usually presented as a pair of graphs of efficiency versus wavelength, one for each polarization. These usually correspond to the grating used in the Littrow or near-Littrow configuration, although the conditions under which the measurements are taken are not always stated. It often tends to be assumed that these curves show *the* efficiency of the grating at a given wavelength. In fact, as we showed in Equation (2.18), the wavelength for which a saw-tooth grating is blazed will depend upon the angle of incidence. Clearly, the efficiency of a grating in a given order changes with the angle of incidence, and therefore a complete specification of efficiency requires measurements to be made at all wavelengths, at all angles of incidence and for light polarized both parallel and perpendicular to the grooves. Indeed some orders disappear "over the horizon" at certain angles of incidence and in Figure 4.16 we showed the range of angles of incidence over which various orders were allowed to propagate. In effect the complete specification of efficiency requires a measurement at each point on that diagram, so that rather than present an efficiency *curve* we need an efficiency *surface* or a three-dimensional graph in which the X and Y axes correspond to the wavelength (or the normalized wavelength λ/d) and the angle of incidence respectively, and the Z axis corresponds to the efficiency (Hutley 1973, Hutley and Bird 1973). Figure 5.23 shows models of the measured efficiency surfaces for a fairly coarse interference grating with a sinusoidal profile. That these surfaces are far from flat indicates the necessity of specifying the angle of incidence under which efficiency measurements are made. In Figure 4.16(b) we plotted curves corresponding to various instrumental configurations. Since efficiency curves are usually measured in a particular instrument, the curve itself is really a cross-section of the efficiency surface cut along the line corresponding to that particular instrument. It is also interesting to note that one can derive useful information about the efficiency surface by measuring other cross-sections which do not correspond to common instrumental configurations. In particular one can measure at a single wavelength the efficiency as a function of angle of incidence. This corresponds to a cross-section perpendicular to the X (or λ/d) axis, and it is from a series of such cross-sections that the surfaces shown in Figure 5.23 were built up.

The rapid undulations in the efficiency surface, which are particularly prominent in Figure 5.23(a), are known as anomalies and

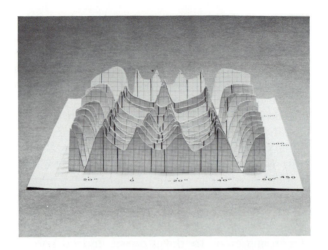

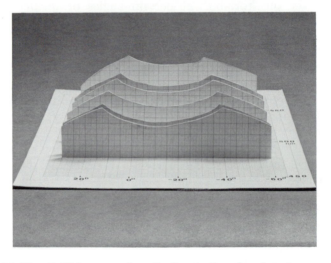

Figure 5.23 The "efficiency surfaces" of a grating showing strong anomalies: (*a*) *S* polarization; (*b*) *P* polarization.

are associated with complex interactions between the electromagnetic fields of the incident and diffracted waves and the surface of the grating. They were first discovered in 1902 by R. W. Wood and are known as Wood's anomalies. We shall discuss them in more detail in Chapter 6, but for the moment we must bear their existence in mind when measuring and understanding the efficiency of gratings. Not all gratings have efficiency surfaces which are as complex as the ones shown in Figure 5.23. Indeed these were chosen as particularly bad examples in order to illustrate the point. With ruled gratings one has some measure of control over the anomalies by choosing the angles of the facets of the grooves, whereas with sinusoidal gratings one is forced to work in an area which is free from anomalies. However, it should never be assumed that any grating is free from anomalies until some fairly detailed measurements of the efficiency surface have been made. The rapidity with which the efficiency can vary near an anomaly demands that one should be particularly careful when extrapolating between two measurements. For example, take the Littrow efficiency curve shown in Figure 5.24. The solid curve shows the detailed measurement of efficiency for the E vector perpendicular to the grooves (the so called S polarization) and the dotted curve shows the result that would have been obtained by extrapolating between measurements made every 50 nm.

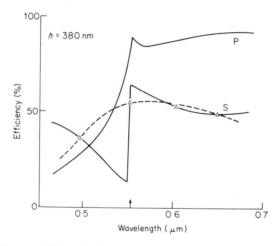

Figure 5.24 The possibility of missing an anomaly if too few efficiency measurements are taken.

The plotting of Figure 5.23 entailed taking some 2000 individual measurements of efficiency. The more complex a surface is, the more measurements are required, but even with a well behaved grating it is impractical to measure the full efficiency surface except for esoteric reasons. In practice one measures an efficiency curve either by using some instrument which gives a continuous scan of the spectrum or by taking a series of individual measurements at discrete wavelengths.

If the grating is to be used in an instrument which differs significantly from the one in which the measurements are made then the efficiency curve may well be different, but the relationship between these two curves can be more readily understood if both are thought of as different cross-sections of the same efficiency surface.

The direct measurement of absolute efficiency is in principle very simple. One merely measures the flux present in the incident beam and then, preferably using the same detector, measures the flux in the diffracted beam. The only problems that arise are the stability of the light source and the linearity of the detector, both of which may readily be checked. One seldom requires a very high degree of photometric accuracy in these measurements as one is usually content to know that a grating is 60% efficient rather than 50% and the distinction between, say, 61% and 60% is of little practical significance. This type of measurement can easily be performed by shining a laser beam onto the grating provided of course that there is available a laser giving a suitable wavelength. The krypton ion laser is particularly useful in this case because it gives a variety of lines covering the visible spectrum and extending to 799 nm in the infrared and 351 nm in the ultraviolet. Alternatively, a continuous wave dye laser will give a complete coverage from 400 nm to 1 μm. The advantage of using a laser is that it is simple and that the beam occupies only a small area so it is well suited to spot measurements on the grating surface. However, one usually wishes to know the efficiency of the grating as a whole, so the beam must be expanded to fill the aperture. It then becomes necessary to take into account the reflectance of the optics used to expand the incident beam and the optics used to focus the diffracted beam down onto the detector. This problem is overcome most simply by comparing the signal detected from the diffracted beam with that obtained when the grating is replaced by a mirror of known reflectance. In this way the losses due to the collimating and focusing optics are common to both measurements and their effects cancel. If S_g is the signal measured with the grating and S_m is the signal measured with a mirror of reflectance R_m, then the absolute efficiency of the grating is

$$\frac{S_g}{S_m} R_m$$

If the mirror is of the same material as the grating then the relative or groove efficiency is simply S_g / S_m.

Most of the metals with which gratings are coated, in order to give as high a reflectance as possible, deteriorate upon exposure to the atmosphere with a consequent loss in reflectance. If, therefore, one wishes to avoid having continually to recalibrate the reference mirror it is important to choose a reference material which is very stable. At the National Physical Laboratory the reference mirror is aluminium vacuum-deposited into a silica flat, but it is the inside surface of the aluminium that is used: i.e. the silica–aluminium interface. The incident beam passes twice through the silica, but this is of no consequence since the reflectance of the mirror was measured under these conditions. The reflecting surface is thus protected from atmospheric attack by the silica on one side and by the bulk of the aluminium on the other. Figure 5.25 shows a typical apparatus for the measurement of efficiencies.

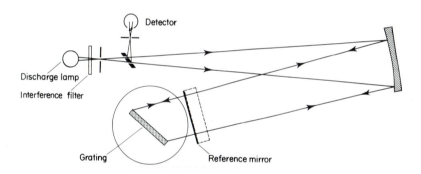

Figure 5.25 Apparatus for the measurement of grating efficiency by comparison with a calibrated reference mirror.

The source may be a discharge lamp which provides a series of single spectral lines, one of which is selected with an interference filter, it may be a thermal source which emits a continuum from which a monochromator selects a narrow band, or it may be a laser. If a single line is selected from a discharge lamp with an interference filter, then it is very important to ensure that the "leakage" of other wavelengths is negligible. Any light of a different wavelength which enters the system will be detected when it is reflected by the mirror, but will be dispersed away from the detector by the grating. This will lead to pessimistic values of the grating efficiency. In order to avoid this a small monochromator may be used in place of the filter. If the incident light is provided by a broad-band source used in conjunction with a monochromator, then stray light from the monochromator will have the same effect as leakage through an interference filter. Although in many cases a high degree of accuracy is not required, this source of error should not be forgotten. The monochromator presents the grating with a band of spectrum of which the bandwidth is determined by the dispersion of the monochromator and the width of the slits. The grating under test may either disperse this short length of spectrum still further or, depending upon its orientation, it may counteract the dispersion of the monochromator. It would do one for positive orders and the other for negative orders. It is therefore necessary to ensure that the exit slit and detector are large enough to collect all of the diffracted light from the gratings. There is a danger, particularly if the monochromator has low dispersion and the grating to be tested is highly dispersing, that pessimistic values may be obtained because some of the diffracted light is dispersed away from the detector.

All that has been said so far applies in principle to all wavelength ranges but in particular to the visible and the near ultraviolet regions. In the infrared there are regions for which suitable detectors are not readily available and it is sometimes necessary to purge the apparatus with dry nitrogen in order to avoid absorption of the radiation by atmospheric water vapour. One has to distinguish between the radiation from the source and the thermal radiation from the walls of the apparatus, and alignment is made more difficult because IR radiation is invisible. Similarly, in the ultraviolet for wavelengths less than about 200 nm it becomes necessary either to purge or to evacuate the apparatus and there are also problems with detectors, sources and the low efficiency of optical components. However, both in the infrared and the ultraviolet these are problems which are associated with spectroscopy as a whole in these regions and not specifically with the measurement of gratings.

In the ultraviolet, for wavelengths ranging from about 200 nm down to X-rays at 0.1 nm or less, the measurement of efficiency is particularly cumbersome. This is mainly because the whole of the apparatus has to be enclosed in a vacuum tank. Also, because of the variability of detectors and the susceptibility to contamination of reference mirrors, the only means of achieving satisfactory results is to use the same detector to measure the incident and diffracted beams and the detector has to be moved about inside the vacuum chamber during the course of the measurements. In practice, therefore, it is necessary to build a special apparatus rather than to use an existing spectrometer. Since this will have to cope not only with plane gratings but, more usually, with concave gratings of various

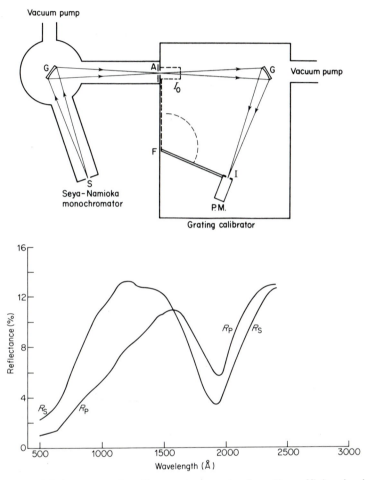

Figure 5.26 (*a*) Apparatus for the measurement of grating efficiencies in the vacuum ultraviolet and polarized efficiency curve (courtesy of E. T. Arakawa). (*b*) Typical result showing the behaviour of Woods anomalies of a UV grating.

radii of curvature, the design of this appartus is likely to be rather complex.

To illustrate this point we show in Figures 5.26, 5.27 and 5.28, three pieces of equipment which have been built specifically for the measurement of grating efficiencies in the vacuum ultraviolet or X-ray regions. The first (Figure 5.26(*a*)) was built at Oak Ridge National Laboratory, Tennessee, USA (Hammer *et al.* 1964). It consists basically of a Seya–Namioka monochromator in which the concave grating under test focuses radiation from an entrance slit on

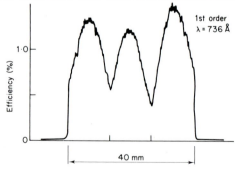

Figure 5.27 (*a*) Apparatus for the measurement of the efficiency of concave gratings across the grating aperture; (*b*) the curved track upon which the grating is mounted; (*c*) typical efficiency map of a tripartite grating. (Courtesy of W. R. Hunter.)

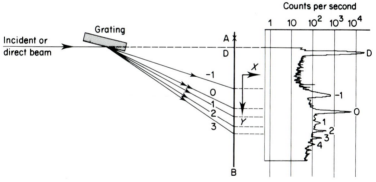

Figure 5.28 Apparatus for the measurement of X-ray grating efficiency. (Courtesy of R. J. Speer.)

to a detector. The instrument is illuminated by monochromatic radiation from a separate monochromator, and the detector is able to swing round to measure the flux passing through the entrance slit and then back to measure that in the focused diffracted order. In this way polarized measurements of efficiency were obtained for wavelengths between 30 nm and 600 nm. Figure 5.26(*b*) shows a typical result which shows the behaviour of Woods anomalies on an ultraviolet grating (Hanson and Arakawa 1966).

The apparatus shown in Figure 5.27(*a*) was built at the Naval Research Laboratories, Washington DC (Hunter *et al.* 1971). The input to the test chamber is provided by a suitable monochromator and source depending upon the wavelength region of interest. For wavelengths down to about 100 nm (for which a suitable window material can be found) the monochromator is used in conjunction with a variety of discharge lamps; for shorter wavelengths an RF helium discharge is employed. The instrument is capable of measuring efficiencies for wavelengths down to 20 nm. The detector is mounted on an arm which enables it to measure both the incident and the diffracted beam and by inserting appropriate extra sections to the vacuum tank gratings of up to 5 m radius of curvature may be measured. One unique feature of this apparatus is that the grating is mounted on a circular track (Figure 5.27(*b*)) with a curvature matching that of the grating blank. In this way the grating can be moved sideways without disturbing the alignment and it is possible to scan the efficiency at different points of the grating surface. This is particularly important for ruled concave gratings, since in manufacture the facet angle can only be set to be correct at one point on the grating surface. Owing to the curvature of the blank, the blaze angle becomes progressively in error and the efficiency falls. Figure 5.27(*c*) shows a scan of the efficiency of a concave grating which was ruled in three sections with three different settings of the diamond in order to alleviate this problem.

Finally, the apparatus shown in Figure 5.28 was constructed at Imperial College, London (Speer 1970) for the measurement of the efficiency of gratings used at glancing angles of incidence for X-ray wavelengths between 15 nm and 0.5 nm. The source, the grating and the detector are all mounted on moveable carriages which are independently controlled so that gratings of various radii of curvature may be tested in the different diffracted orders at different angles of incidence. The whole mechanism is mounted in a vacuum tank $3\,\text{m} \times 2\,\text{m} \times \frac{1}{2}\,\text{m}$ and is computer-controlled. The special problems associated with X-ray gratings will be discussed further in Chapter 8, but for the moment it is appropriate to note the impressive amount of engineering that is required to put into practice what is basically a very simple measurement.

6

Anomalies and the Electromagnetic Theory of Grating Efficiencies

□   Anomalies and the theory of efficiency

In an ideal case one might expect that the efficiency of a grating would vary smoothly from one wavelength to another. In practice there are often localized troughs or ridges in the efficiency surface and these are observed as rapid variations of efficiency with a small change of either wavelength or angle of incidence. They were first observed by R. W. Wood in 1902 as a series of light and dark bands in the spectrum from a white light source; a typical example of the phenomenon is shown in Figure 6.1. Anomalies are strongly dependent upon the polarization of the incident light and are most prominent when the electric vector is perpendicular to the grooves − the so called S polarization (S from German *Senkrecht*, perpendicular) − and this feature is brought out in Figure 5.22. They may be observed in the orthogonal P polarization (i.e. parallel to grooves) but usually only with gratings having deep grooves (Palmer 1952). From the point of view of the spectroscopist, anomalies are a nuisance because they introduce spurious peaks and troughs into an observed spectrum. In the same way that a ghost might be mistaken for an emission line, an anomaly can easily be confused with an absorption band. It is, therefore, very important that the positions and shapes of the anomalies should be known before an instrument is used. Some instruments, such as double-beam spectrophotometers, are less susceptible than others to the effects of anomalies, but it is nevertheless important to be aware of their existence. The extreme suddenness with which the efficiency can vary was demonstrated

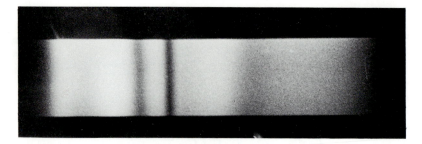

Figure 6.1 A typical Wood's anomaly.

by R. W. Wood who effectively extinguished one of the sodium D lines but not the other.

However, anomalies do serve a useful purpose when one attempts to compare theoretical predictions of efficiency with the measured values. They provide the efficiency curves with detail which is a very sensitive test of the accuracy of any theory. Once an adequate theory is established it is possible to work back from the experimental observation of anomalies and determine some features of the nature of the grating surface.

In the present chapter we shall consider anomalies and some of the more detailed theories which may be used to describe the efficiency of a grating in a given order. In Chapter 2 we described the relationship between the various aspects of a grating's performance in terms of the Fourier transform of the grating itself. This was quite adequate when describing resolution, stray light and ghosts, but we warned that in general it was inadequate to describe the distribution of the incident light among the various orders (the envelope function in Figure 2.25) as a simple Fourier transform of the groove profile. Let us now consider what must take its place. While most of the grating theory required for practical applications was well understood by Fraunhofer in the mid-nineteenth century, the question of a satisfactory theory for the distribution of light among the diffracted orders of an optical grating was not adequately resolved until the early 1970s. It is, therefore, only comparatively recently that it has been possible to predict the observed efficiency of a grating from the shape of the groove profile. Even this has only been possible with the aid of powerful computers and considerable skill in formulating the problem in such a way that it is amenable to numerical analysis. The mathematical manipulations that are required for the solution of the general diffraction grating problem are beyond the scope of this book. We shall, however, examine various theories and consider the physical models on which they are based, the assumptions that are made and the ranges of application over which the theories are valid.

The first and simplest explanation of anomalies was given by Rayleigh in 1907, who suggested that they occur when an order "passes off over the grating horizon". That is, when the angle of diffraction is 90° and the order skims the grating surface. For shorter wavelengths the order propagates in the normal way and for longer wavelengths no diffracted order is possible. Thus at the wavelength which is grazing the surface, the so-called Rayleigh wavelength, there is a discontinuity in the number of orders that are allowed to propagate. The energy that is in the order which passes off has to be redistributed among the other orders and this accounts for the sudden fluctuations of the efficiency of these orders. This explanation was successful in describing the position of many of the anomalies described by Wood to such an extent that Rayleigh was able to suggest correctly that on one occasion Wood had mistaken the value of the pitch of one of his gratings. However, this explanation was not sufficient to explain the position of all observed Wood's anomalies; it was further demonstrated by Strong in 1936 that in some cases the position of the anomaly depends upon the material of which the grating surface was made. This could clearly not be described by Rayleigh's theory because the passing off of an order depends simply upon the wavelength of light, the angle of incidence and the pitch of the grating. It was suggested that those anomalies which did not fit the Rayleigh explanation might be due to some resonance effect within the grooves themselves. Palmer and Phelps (1968) demonstrated in the microwave region that anomalies could be observed with as few as three grooves. Unfortunately the resolution of a grating with few grooves is very small, so although the anomaly may be there it becomes very difficult to observe because its effect is swamped by diffraction.

The condition that a diffracted order shall graze the surface is that the angle of diffraction shall be 90°. This was discussed in Chapter 4 and in Figure 4.16 we showed a graph of the angle of incidence against the value of λ/d for which various diffracted orders pass off. The similarity between this curve and the experimental measurements of the efficiency surface in Figure 23 of Chapter 5 is quite evident. The positions corresponding to the "Rayleigh angle" are marked as vertical lines on the individual efficiency curves which make up the model of the efficiency surface and it is interesting to note that the features which correspond to the Rayleigh theory are either small cusps or sudden changes of slope. Although the most prominent features do follow the same general pattern, they are displaced from other anomalies. It should perhaps be stressed at this stage that the grating from which Figure 5.23 was derived had been chosen especially because it showed strong anomalies; it should not be taken as typical of either ruled or interference gratings as a whole. It is, however, worth noting that the conditions under which anomalies may be completely avoided are just the same as those under which reasonably high efficiencies may be attained from gratings with a symmetrical profile. That is, that only one diffracted order should be permitted to propagate.

One feature of anomalies that has been known ever since Wood discovered them is that they are far more pronounced for light which is polarized with the electric vector perpendicular to the grooves than for light polarized parallel to the grooves. There is a very simple explanation for this in terms of Rayleigh's hypothesis. For the P polarization the electric field of the order which is passing off is parallel to the surface of the grating, whereas for the S polarization it is perpendicular, as shown in Figure 6.2. If the grating is a good electrical conductor, then it will not sustain an electric field in the surface, so the field strength in a P polarized order propagating parallel to the surface is much less than that for an S polarized order. Therefore, there is less energy to be redistributed when the order passes off, and hence the anomalies are less severe.

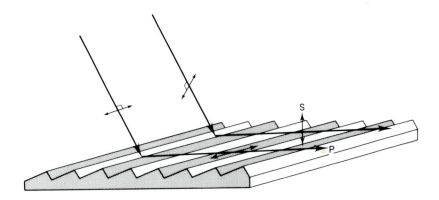

Figure 6.2 The propagation of waves along the surface of a grating.

☐   Grating theories

The test of any theory is that it shall accurately predict the values
of efficiency that are measured in practice. However, the confron-
tation of theory and experiment is not always the best means of
testing a theory. It may happen that a good agreement is found by
luck, if one happens to choose a grating for which the efficiency curve
is insensitive to variations in the groove profile or to the inaccuracies
of the theory. It is not always possible to measure the efficiency in
all desired geometries, nor is it always possible to measure the groove
profile with adequate accuracy. For many years it was believed that
the inadequacy of measurements of groove profile for very fine gratings
was responsible for the fact that there was often poor agreement
between theory and experiment in the visible region (e.g. Stroke 1967),
whereas good agreement was achieved when the same experiment
was scaled up to the microwave region where the grating grooves
could be measured easily (Deleuil 1969). We now know that this was
not the correct explanation, but it does indicate the need for more
than just a reasonable agreement between theory and experiment
before the validity of a theory can be established. The first test that
should be applied to any theory is that it should be self-consistent.

Until the early 1970s it was assumed almost universally that the
grating could be treated as though the surface was perfectly reflecting
and therefore infinitely conducting, and that in order to correct the
results for the effects of finite conductivity it would be necessary
merely to multiply them by the observed value of the reflectance.
Under these circumstances the energy absorbed by the grating must
be equal to that absorbed by a (thick) mirror of the same material
and the total energy in the diffracted orders must be equal to the
incident energy minus that absorbed by such a mirror. In other words,
the sum of the efficiencies in the different orders should equal the
reflectance of the material. We now know that this assumption is not
always valid and that in practice the effects of finite conductivity
must be considered from the outset. Nevertheless, if a theory gives
inconsistent results when applied to an infinitely conducting grating
it can certainly not be applied to the case of a real grating.

The calculation of grating efficiency demands a great deal of
computation and it is only with the help of electronic computers that
it has been possible to develop the theory to the stage where it can
be applied with confidence to a wide range of physical situations. We
now consider the development of various theories that have been put
forward.

For the sake of completeness let us start by considering the scalar theory of a perfectly conducting grating. Even though we know in advance that its usefulness is limited because it takes no account of polarization, it is still instructive to consider its self-consistency. The scalar theory adopts the approach outlined in Chapter 2 in which the contributions to the amplitude of a wave at a given point are summed for components coming via all points on the grating surface. If the source and points of observation are both considered to be at infinity, then in effect we calculate the Fourier transform of the grating. As we saw in Chapter 2, the nature of the Fourier transform enables us to consider independently the effects of the finite aperture, the periodic nature of the grating and the shape of the grooves so that it is only necessary to carry the integration over one groove.

This theory was put into a general form by Madden and Strong (1958), who expressed the disturbance dE_P at point P due to a Huygens wavelet emitted from an elemental area of grating surface ds in the following form (Kirchoff's differential):

$$dE_P = \frac{iA}{2\pi r'' r'} \exp\left\{i\omega\left(t - \frac{r'' + r'}{c}\right)\right\} \left[-\cos(n \cdot r'') - \cos(n \cdot r')\right] ds$$

(6.1)

where, as shown in Figure 6.3, n is the unit vector normal to the surface element ds, r'' and r' are the vector distances from the source Q and point of observation P respectively, and A is the amplitude of the scalar wave field at the point source Q. In order to solve the problem completely, it would be necessary to integrate this expression over the whole of the grating surface for all points P. However, because of the simplifications which arise from the Fourier transform relation between the grating surface and the spectral image plane, it is only necessary in practice to integrate over one groove for directions of r'' which are given by the grating equation.

Let us now consider a fairly typical blazed grating having facet angles of $10°$ and $80°$ and calculate the efficiencies in all diffracted orders for an angle of incidence of $20°$ for a range of wavelengths. The actual value of wavelength is immaterial in this case and it is the "normalized" wavelength λ/d which is important. If the theory is self-consistent, then we would expect the sum of efficiencies to equal unity. The results are shown in Figure 6.4. The total varies from 0.48 to 0.94 and in fact for other angles of incidence the variations are even more extreme (McPhedran 1973).

The rather sudden features that arise in this graph correspond to the passing off of various diffracted orders. It is not surprising

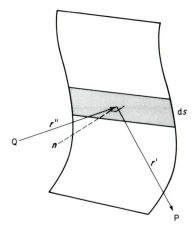

Figure 6.3 Radiation from an element of grating surface.

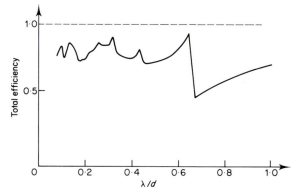

Figure 6.4 The total efficiency calculated according to scalar theory of a grating with a triangular profile with facet angles of $10°$ and $80°$. (After McPhedran 1973.)

that these oscillations occur, because the theory does not impose any boundary condition that would couple together the various orders. Any one order has no "knowledge" of what is happening in any other order and therefore does not accommodate the energy which has to be redistributed when an order passes off. There are regions in which the scalar theory can be most useful and it will no doubt remain the basis for our intuitive understanding of diffraction grating phenomena, even in regions where it does not apply. As a general rule, it is often accepted that scalar theory can be applied when the groove spacing is greater than five times the wavelength. This is based on the observation that, anomalies apart, polarization effects are not very serious for $\lambda/d < 0.2$. A more detailed survey of the results of the theory reveals that it is inconsistent even when the groove spacing is ten times the wavelength (Cerutti–Maori and Petit 1970), particularly with deeper grooves and higher angles of incidence. This may partly be accounted for by the fact that in the form described by Madden and Strong and studied by McPhedran (1973), the theory takes no account of multiple scattering; that is, light which is scattered from one facet to another before leaving the grating. Various attempts have been made to include these effects (e.g. Janot and Hadni 1962, Palmer and LeBrun 1972), but none has enabled the scalar theory to give an adequate quantitative description of grating efficiency in all domains.

A rather more rigorous theory had been put forward by Rayleigh in 1907, in which the various diffracted orders were coupled together by a boundary condition on the surface of the grating. Rayleigh expressed the field strength $\bar{E}_d$ in the region above the grating as the sum of the outgoing plane waves of amplitude B_n. Far away from the grating the field consists of a series of outgoing plane waves of constant amplitude, whereas in the immediate vicinity of the grating it was necessary to take account also of evanescent waves corresponding to the case where $|\sin\theta| > 1$ and where the amplitude decreases exponentially with distance from the surface of the grating.

If we consider, for simplicity, the case of a P polarized wave incident on a perfectly conducting grating of period d and a groove profile represented by $y = f(x)$ as shown in Figure 6.5, then the z component of the diffracted field may be represented by

$$\bar{E}_d = \sum_{-\infty}^{\infty} B_n \bar{E}_n = \sum_{-\infty}^{\infty} B_n \exp\left[ik(x\sin\beta_n + y\cos\beta_n)\right]\hat{z} \quad (6.2)$$

where β_n is the angle of diffraction of the nth order. If $\bar{E}_i$ is the incident field strength then the boundary condition *on the surface of the grating* is simply that $\bar{E}_d + \bar{E}_i = 0$, or

$$-\bar{E}_i = \sum_{-\infty}^{\infty} B_n \bar{E}_n(x, f(x)) \quad (6.3)$$

$\bar{E}_i$ and $\bar{E}_n$ are known, and we wish to calculate the unknown quantities B_n which are the amplitudes of the various diffracted orders, so the mathematical problem is one of finding the coefficients in the expansion of a known function in terms of an infinite set of

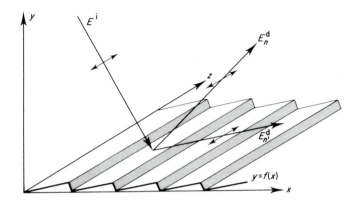

Figure 6.5 Diffracted and incident fields according to the Rayleigh theory.

functions. Various solutions to this problem have been proposed. Some are more appropriate than others to different forms of groove profile and further details of these may be found in the works of Meecham (1956), Stroke (1960) and Petit (1963, 1965). For the present we are concerned not with the detailed mathematics but with the extent to which the theory is self-consistent. It is again possible to calculate the total diffraction efficiency from an infinitely conducting grating and the results for the blazed grating with $10°$ and $80°$ facet angles are presented in Figure 6.6. These may be compared directly with those of Figure 6.4 for the scalar theory.

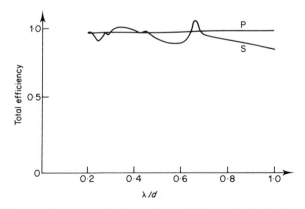

Figure 6.6 Total efficiency of the grating of Figure 6.4 calculated according to the Rayleigh theory. (After McPhedran 1973.)

The scalar theory gave values of total diffracted energy lying between 0.48 and 0.94. For P polarization, the Rayleigh theory yields values of between 0.96 and 0.99, but the S polarization is less well behaved and the total varies between 0.86 and 1.06. This is a significant improvement over the scalar theory and, because the various orders are coupled together by the boundary condition at the grating surface, the Rayleigh theory does predict the anomalies that occur when an order passes off.

There is, however, in this theory one basic assumption that is not really justified. It is assumed that the field at any point above the grating surface may be represented by the sum of the incident, diffracted and evanescent orders; although it is true outside the grating it is not necessarily true in the space within the grating grooves, that is, the region $y < f(x)_{\max}$ in Figure 6.7. The effect of the incident field will be to induce circulating currents at the surface of the grating and these in their turn will generate travelling electromagnetic waves.

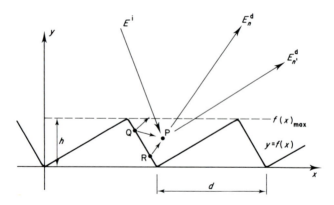

Figure 6.7 Contributions to the electric field within a deep groove.

If we consider a point P within the groove, it will receive contributions to the field from the incident wave and the diffracted waves such as those due to radiation out of the grating from point R. It will, however, also receive a contribution from waves travelling into the grating from points such as Q and it is this contribution that is neglected in the Rayleigh expansion. It is intuitively evident that this omission becomes more serious the deeper the grooves of the grating and rigorous theoretical studies (Petit and Cadilhac 1966) have indicated that in general the theory is valid provided that the depth is less than about 7% of the groove spacing ($h = f(x)_{max} < 0.07d$ in Figure 6.7). However, even this result depends upon the form of the grooves and in particular upon whether or not the grooves have sharp discontinuities (Millar 1969, 1971). These comments give some physical insight into why Rayleigh's theory often gives unsatisfactory results. From a mathematical point of view, its validity has been the subject of some debate and concerns questions of the nature of the convergence of infinite series. Such matters are, however, beyond the scope of our present discussion (Wirgin 1980).

A more rigorous approach that took account of the effects of forward scattering within the grating grooves was put forward by Pavageau and Bousquet in 1970. They sought first to calculate the density of surface current induced on a perfectly conducting cylindrical surface of directrix C by an incident magnetic field H_i. This they expressed as Fredholm integral equation,

$$\phi(x) = \phi_0(x) + \frac{1}{d} \int_0^\infty N(x, t)\phi \, dt \qquad (6.4)$$

where $\phi(x)$ is the induced surface current, $\phi_0(x)$ is that due to the incident field and $N(x, t)$ is a kernel function which can be evaluated numerically. It is then possible to calculate the intensity of the diffracted wave from the expression for the retarded vector potential $A(p)$ which may be written

$$A(p) = \frac{-i\mu_c}{4} \int_C \phi(m)H_0^2(k_p, M) \, dm \qquad (6.5)$$

where H_0^2 is the Hankel function of the second kind and order zero.

This formalism gives results which are dramatically more self-consistent than those of the Rayleigh expansion. In Figure 6.8 we see that the total diffracted energy for our test example is very close to unity for both states of polarization for a wide range of values of λ/d. We may therefore assume from the purely theoretical standpoint that this formalism is adequate to describe the diffraction of light by a grating and we can go on to consider how well in practice the theory is able to predict the measured efficiencies of real gratings.

All that has been said so far has been based on the assumption that there is no loss of accuracy by assuming that the grating is perfectly reflecting and then modifying the results by the observed value of the reflectance of the metal of the grating surface. This

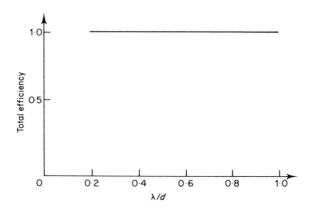

Figure 6.8 Total efficiency of the grating of Figure 6.4 calculated according to the theory of Pavageau and Bousquet. (After McPhedran 1973.)

assumption greatly simplifies the mathematics and is acceptable in regions which are well away from anomalies and for long-wavelength radiations (millimetre waves to far infrared), where the conductivity of good metals is very high. Excellent agreement was obtained, for example, by Deleuil between measurements made on large gratings with microwaves and the theoretical prediction based on the Pavageau and Bousquet formulation. However, when a similar comparison was made in the infrared, the agreement between theory and practice was very much worse. It was not possible on this occasion to repeat the experiment with visible radiation, but there is little doubt that the agreement would have been worse still. Because of the difficulties of measuring the groove profile with any accuracy, it was some time before the cause of the disagreement could be attributed unambiguously to the shortcomings of the theory rather than to the inadequacy of the experiment.

There were two experimental phenomena that indicated that the infinite conductivity theory was inadequate. First the demonstration by Strong in 1936 that the *position* of some types of anomaly depended upon the material with which the grating was coated. If, as was assumed, it was sufficient to correct for the finite conductivity of the metal by multiplying the result for a perfectly conducting grating by the observed reflectance of that metal, then any anomalies would have stayed in the same position.

Strong applied a series of different metal coatings to several gratings and observed that the positions of some anomalies were independent of the nature of the coating while others shifted quite significantly. Those that stayed the same obeyed the conditions predicted by Rayleigh in terms of the passing off of a higher order. The other anomalies that were usually less sharp but more intense, did not fit into Rayleigh's explanation but they appeared to be linked to those which did. In 1965 Hessel and Oliner suggested that there are in fact two distinct types of anomaly. Those that we might call Rayleigh anomalies are due to the passing off of higher diffracted orders, whereas the others are due to a resonance phenomenon.

The second phenomenon was the observation reported by Häglund and Sellberg in 1966 that when the sum of the diffracted orders is measured using an integrating sphere it does not equal the reflectance of the metal or even remain constant but varies dramatically in the region of an anomaly. In some cases a significant fraction of the incident light fails to leave the grating at all. It is obvious that the criterion that we have been using to assess the self-consistency of the different theoretical formulisms has little relevance in practice. It remains a valid test of any analysis that, if the material is assumed to be infinitely conducting, then the total diffracted energy should equal the incident energy; this can be used to check both the validity of the way in which the problem is formulated and the accuracy of the mathematics. It does not, however, necessarily relate to reality. Even the formulism of Pavageau and Bousquet, which was theoretically very sound and which had successfully predicted grating efficiencies measured at long (mm) wavelengths, was inadequate to describe the anomalies of gratings measured in the visible. This was confirmed by a very detailed comparison between a set of theoretical curves calculated by McPhedran and Waterworth (1972) for gratings with a sinusoidal profile and a corresponding series of measurements made by Hutley (1973) on interference gratings. An example of the results is shown in Figure 6.9 and from this we see that there is an excellent *quali-tative* agreement between both curves. In many cases the anomaly consists of a sharp peak at the Rayleigh wavelength followed by a trough and this is shown in both the experimental and theoretical curves. There is, however, a very striking discrepancy in the wavelength scales. The theory predicts that the separation between peak and trough should be of the order of 0.5 nm, whereas the measured separation is over 50 nm; different by a factor of over a hundred.

A similar result was obtained on entirely theoretical grounds by Maystre (1973, 1974). He adapted the theory of Pavageau and

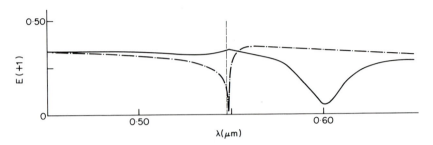

Figure 6.9 The discrepancy between the observed positions of anomalies and those calculated assuming a surface of infinite conductivity.

Bousquet to take account of the finite conductivity of the surface by incorporating measured values of the optical constants of metals. He showed that for P polarization it was sufficient to multiply the results obtained with infinite conductivity by the reflectance of the metal. On the other hand, for S polarization the finite conductivity was responsible for substantial changes in the positions of some anomalies. He also applied this theory to the case of sinusoidal gratings used in the conditions studied by McPhedran and Waterworth and by Hutley and established that the discrepancy was due solely to the finite conductivity of the metal. In other respects the agreement between theory and experiment was not particularly good and this may have been due to deficiencies in the experimental data.

In order to confirm that the finite conductivity theory was capable of accurately predicting the efficiency of real gratings in the visible, a comparison was made between a very detailed experimental study by Hutley and Bird (1973) of the efficiency of a sinusoidal grating, at various wavelengths for different angles of incidence and coated with different materials, and the calculations performed by McPhedran and Maystre (1974) at Marseille with the finite conductivity theory using experimental data supplied by Hunter of NRL Washington for the optical constants of aluminium, gold and silver. The agreement was excellent and two important sets of results are reproduced in Figures 6.10 and 6.11.

Figure 6.10 shows for a single wavelength the efficiency in all diffracted orders for all angles of incidence and reveals several interesting features. It enables us to see for any given angle the relationship between the light in the various diffracted orders. For example, at an angle of incidence of about $-20°$, we see a dramatic fall in the efficiency in the $+1$ order and a corresponding increase in efficiency in the zero and -1 orders. It would therefore appear that there is strong coupling between the orders which enables energy to be transferred out of the $+1$ order and redistributed among the -1 and 0 orders. Furthermore, this process is lossy because there is also a dip in the total efficiency curve and we must therefore assume that the missing energy is absorbed by the grating. Another interesting feature is the symmetry that each curve possesses with respect to its own Littrow angle. In effect this means that for a given angle of incidence, the direction of the incident and diffracted beams may be reversed without altering the efficiency. This result is an example of the Helmholz reciprocity theorem. It has been justified for gratings on theoretical grounds both for finite and infinite conductivity (Maystre and McPhedran 1974) and there is considerable experimental evidence of its validity over a wide range of applications for both sinusoidal and blazed groove profiles. (There is sometimes confusion, however, concerning the validity of the theorem in the case of echelles where the grooves are vignetted. This question is discussed briefly in Appendix 6.1.)

Figure 6.11 shows the variation of efficiency of a sinusoidal grating at a single wavelength as a function of angle of incidence. The curves for the -1 order and total diffracted energy are shown for the case in which the grating is covered by aluminium, silver and gold and this enables us to study in more detail the effect of changing the material of the grating surface. The first feature that we notice is that the curves for silver and aluminium are basically the same shape except that the position of the main features are changed. The sharp peak at the Rayleigh angle of about $-37°$, for example, is fixed, but the position of the adjacent strong minimum varies. At the wavelength at which these measurements were taken (476 nm) gold is strongly absorbing with a normal reflectance of only 40% and we see that in this case the anomalies are barely discernible. This is in accordance with the observation of Strong who also studied copper and magnesium coated gratings. The final curve shows a measurement for gold at a wavelength at which it is a good reflector and we see that it behaves in the same way as aluminium and silver. We may therefore conclude that it is possible to reduce the strength of anomalies by coating the grating with an absorbing material, but in doing so we pay the price of a substantial reduction of overall efficiency. From these results we see that the theory as formulated by Pavageau and Bousquet and extended to the case of gratings of finite conductivity by Maystre

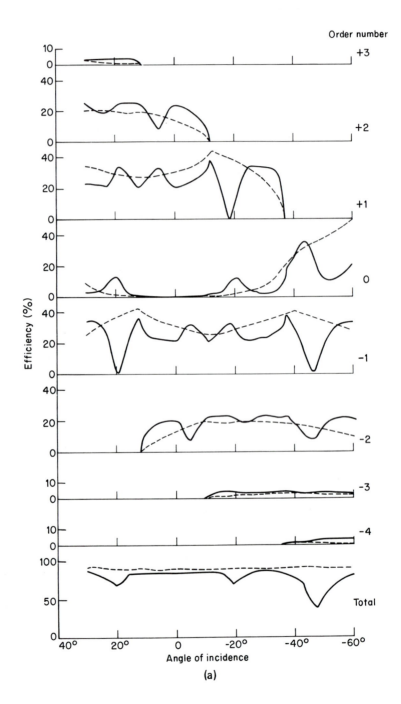

(a)

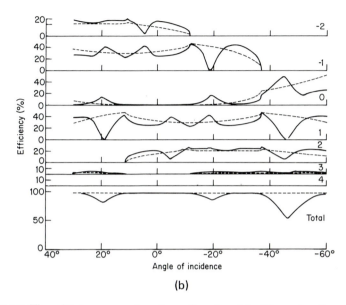

(b)

Figure 6.10 The efficiency as a function of angle of incidence in all orders for a 730 grooves mm^{-1} sinusoidal grating: (*a*) observed data (Hutley and Bird 1973); (*b*) calculated according to finite conductivity theory (McPhedran and Maystre 1974).

is capable of describing all observed details of the performance of a grating used in the visible. Using a somewhat different mathematical approach, Loewen, Nevière and Maystre (1977) have confirmed that this is true for a wide range of grating groove profiles, particularly the saw-tooth profiles which are so widely used in spectroscopy.

Although we can now predict with confidence the efficiency of a grating from a knowledge of the nature and shape of the grooves, this still requires the use of a powerful computer and considerable mathematical dexterity. The theory does not really give us a physical picture of the way in which a grating works and does not tell us which aspects of a grating determine, for example, whether an anomaly will exist at all, where it will be and whether it is likely to be strong or weak. There are, however, other ways of describing anomalies which give us a physical picture that is more easily understood.

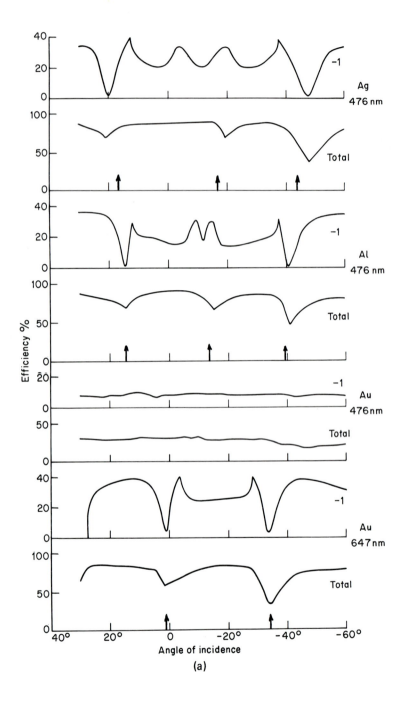

(a)

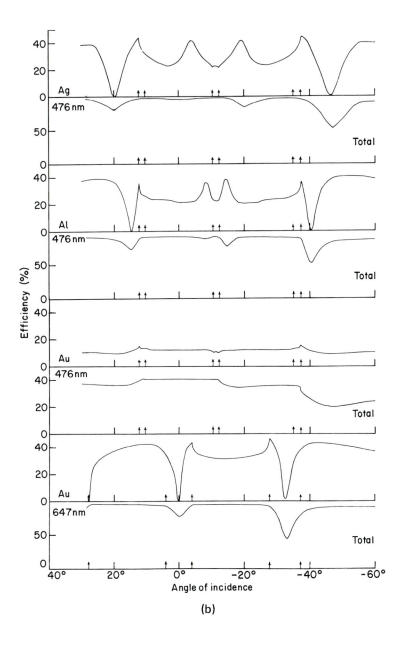

Figure 6.11 First order and total efficiency as a function of angle of incidence for a grating covered with different materials: (*a*) observed data; (*b*) theoretical data.

☐   Surface plasma oscillations

The collective oscillations of the free (conduction) electrons at the surface of a metal can be described as a series of waves. These are longitudinal electromagnetic waves and may loosely be thought of as sound waves propagating in the "free-electron gas" of the metal. These surface plasma oscillations may be thought of either as waves or, often rather more conveniently, in quantum mechanical terms as *surface plasmons*. Their existence was first postulated by Ritchie (1957) who showed that for a given angular frequency ω of the oscillations, the wave vector k is given by

$$k = \frac{c}{\omega}\left(\frac{\epsilon_1(\omega)}{1 + \epsilon_1(\omega)}\right)^{1/2} \tag{6.6}$$

where $\epsilon_1(\omega)$ is the real part of the dielectric function of the metal and is given by

$$\epsilon_1(\omega) = n^2 - k^2$$

where n and k are the real and imaginary parts of the refractive index. It is possible to couple energy into these waves from a magnetic field oscillating parallel to the surface of the metal. If the surface is perfectly smooth then no coupling takes place, but if the surface is rough then there will be local variations in the electric field and the coupling of the energy into the surface plasma oscillations is possible. In particular it is possible to couple energy into surface plasma oscillations from the field associated with electromagnetic radiation which is being reflected by the metal surface. In quantum mechanical terms we can say that a photon is absorbed and a surface plasmon is created in its place. In order that this should happen it is necessary that the tangential component of the wave vector of the incident radiation should match the wave vector of the surface plasmon. For radiation of a given frequency ω (and wave vector ω/c), it is possible to "tune in" to the surface plasmon by adjusting the angle of incidence (see Figure 6.12(a)).

If the wave vector of the surface plasmon is k_{sp} and that of the incident photon is k_p, then the condition for coupling to occur is that $k_p \sin\theta = k_{sp}$. The extent to which the coupling will occur depends upon the roughness of the surface and for most metals the incident radiation must lie in the ultraviolet in order that k_p should be sufficiently large. If one measures the reflectance of a (slightly) rough metal surface in the ultraviolet as a function of angle of incidence, then one can observe an absorption of the type

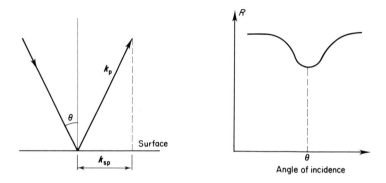

Figure 6.12 The coupling of energy into surface plasmons at reflection.

shown in Figure 6.12(*b*). As one would expect from the theory, this is true only for light with *E* vector parallel to the plane of incidence (i.e. perpendicular to the grooves if the surface were a grating) and the position of the absorption varies with different materials in accordance with the theory (Teng and Stern 1967).

A grating can be regarded as a regular form of roughness, so it is perhaps reasonable to expect that similar effects would be observed on gratings. Furthermore, as we saw in Chapter 2 a grating acts in such a way that it adds to the momentum of a photon a quantum of momentum that is proportional to the groove density. This "grating momentum" can also be taken into account when considering the coupling between the external radiation field and the surface plasmons in the grating, so that our vector diagram takes the form shown in Figure 6.13. We see from this that the condition for coupling is given by

$$k'_p + k_g = k_{sp}$$

or

$$\frac{\omega}{c} \sin \theta + \frac{m 2\pi}{d} = \frac{\omega}{c} \left(\frac{\epsilon_1(\omega)}{1 + \epsilon_1(\omega)} \right)^{1/2}$$

or

$$\sin \theta = \left[\frac{\epsilon_1(\omega)}{1 + \epsilon_1(\omega)} \right]^{1/2} - \frac{m\lambda}{d} \tag{6.7}$$

so that for a given wavelength of incident light and a grating of a known metal it is possible to work out the angles of incidence for which coupling can occur. This results in a reasonable but by no means perfect agreement between the positions of the main features of the constant wavelength efficiency curves and the angles

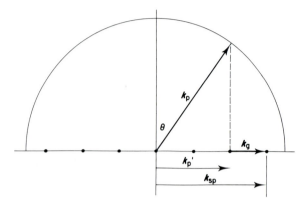

Figure 6.13 The conditions for coupling energy into surface plasmons in a grating.

calculated from Equation (6.7). The agreement is much better than that predicted by the infinite conductivity theory where these features occur practically at the Rayleigh angles. Equation (6.7) is very useful because it enables us to calculate from a knowledge of the grating pitch and the optical constants of the material roughly where we would expect to observe an anomaly. However, it can only be regarded as a first approximation because it takes no account of the depth or the shape of the grooves and we know full well that it is the groove profile that determines the shape of the efficiency surface and that this is nowhere more true than in the region of an anomaly. In deriving this expression we have tacitly assumed that the existence of the grating does not perturb the propagation of the surface plasmons. This is obviously not the case, if only because the surface plasmon will be slowed down on a deep grating because it has to travel further up and down the grooves. Figure 6.14, for example, shows the effect of varying the depth of a sinusoidal grating and demonstrates that as the grooves become deeper there is a perceptible shift in the position of the anomaly.

In order to predict the detailed shape of the efficiency curves in terms of surface plasmon effects it is necessary to calculate the dependence of the excitation of surface plasmons upon the groove profile. Such calculations have been performed and compared most thoroughly with experimental data throughout the visible and ultraviolet spectral regions by Wheeler, Arakawa and Ritchie at the Oak Ridge National Laboratory, Tennessee. It should be emphasized that although the concept of the surface plasmon is rather different

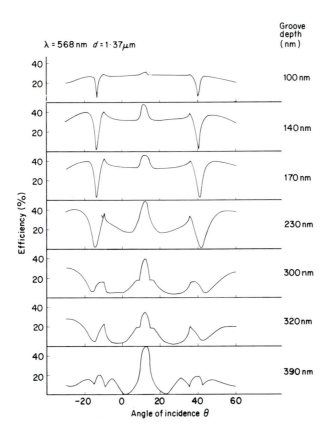

Figure 6.14 The efficiency of a sinusoidal grating as a function of angle of incidence and groove depth.

from the formalism of Pavageau and Bousquet or the Rayleigh expansion, there is no conflict between the surface plasmon description and the classical description of the interaction of light with a corrugated metal surface. The propagation of real and evanescent waves and the propagation of surface plasmons are both governed ultimately by Maxwell's electromagnetic equations. The classical and the surface plasmon approaches are therefore equivalent in that they both describe the same phenomena and are based on the same laws of electromagnetism. However, by introducing the idea of a surface plasmon we do have the advantage of a physical picture of how anomalies occur. It also provides us with a qualitative explanation for another phenomenon which is not so readily incorporated into the classical theories. That is the scattering of light by surface plasmons.

□   ## Scattering from surface plasmons

In the preceding paragraphs, and particularly in Figure 6.13, we described the condition under which an incident photon could be absorbed by the grating in the creation of a surface plasmon. It is important to realize that the reverse process can also take place: a surface plasmon of the appropriate wave vector and travelling in the right direction with respect to the grating can give rise to the emission of a photon. This was demonstrated in 1967 by Teng and Stern who created surface plasmons in the surface of a grating by bombarding it with high-energy electrons at normal incidence. They observed light emanating from the surface of the grating and found that it was concentrated into a series of well defined arcs. This can easily be explained if we consider the case of a surface plasmon travelling in some arbitrary direction in the grating surface, let us say at an angle ϕ to the grooves (Figure 6.15). This plasmon will "see" a grating of spacing $d/\sin \phi$, so the magnitude of the "grating momentum" that takes part in the coupling process will be reduced by a factor of $\sin \phi$ and we can see from Figure 6.13 that a reduction in k_g will lead to an increase in θ.

We may therefore rewrite Equation (6.7) in the more general form

$$\sin \theta = \left[\frac{\epsilon_1(\omega)}{1 + \epsilon_1(\omega)} \right]^{1/2} - \frac{m\lambda}{d} \sin \phi \qquad (6.8)$$

and this governs both the creation of a photon from a surface plasmon and a surface plasmon from a photon. It is also possible for both processes to occur at once, so that incident light may be absorbed as a surface plasmon and then re-radiated in a different direction. In the coupling process the vector k_g is not necessarily determined solely by the periodicity of the grating, but is also influenced by the roughness of the surface of the grooves, and it may not necessarily lie in the direction perpendicular to the grooves. Even with light that is incident in a well defined direction it is therefore possible to generate surface plasmons in all directions in the grating surface and to observe the radiation described by Teng and Stern simply by illuminating a grating with a beam of monochromatic light. This can be seen very effectively by placing a grating in the centre of an opalescent sphere and illuminating it with an unexpanded laser beam, as shown in Figure 6.16.

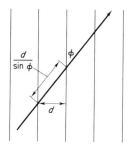

Figure 6.15 The effective pitch of a grating seen by a surface plasmon travelling obliquely to the grooves.

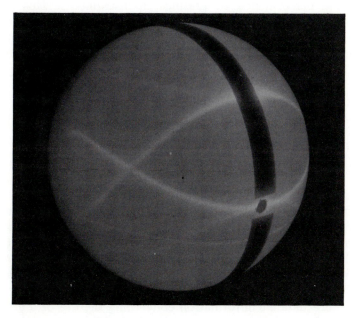

Figure 6.16 Surface plasmon scattering from a grating displayed on an opalescent sphere.

In this particular case the sphere was a domestic light fitting with a small hole to allow the laser beam to enter. In order that the scattering from the diffracted orders should not swamp the surface plasmon scattering, a band of black paper was placed around the equator of the sphere and this absorbed the diffracted orders. The pattern can be detected at all angles of incidence of the beam and this can only be explained in terms of the roughness of the grating contributing to the grating wave vector k_g. However, the scattering is significantly stronger when the angle of incidence is such that Equation (6.8) is satisfied and the coupling is due solely to the periodicity of the grating. As the grating is rotated about an axis parallel to the grooves, the pattern of scattered light remains fixed with respect to the grating. This is to be expected, since the process of creating a photon through interaction of a surface plasmon with the grating does not depend upon the means by which the surface plasmon was itself created. The diffracted orders will rotate through approximately double the angle through which the grating is moved so that their relation with the scattering pattern changes with angle of incidence.

The phenomenon of surface plasmon scattering does not appear to affect the performance of gratings in instruments. This may be because it is rather diffuse and in ruled gratings it would be masked in the plane of the spectrum by "grass". Even with interference gratings it is unlikely that it would ever be detected in an absorption instrument because the contributions from different wavelengths would merge to give a general background level of scattering. It may well be possible to detect it when studying emission spectra, but even then, because of the diffuse nature, it may not matter in practice.

□   Brewster angle effects, the reduction of theoretical
       data and the total absorption of light by a grating

The three topics listed in this heading might at first appear to be totally unconnected, but they are all features of a rather different approach to anomalies that has been put forward by Maystre and Petit (1976). Their aim in doing this was to reduce the enormous amount of data generated by the computer to a few simple parameters which would characterize the performance of a given grating. From these, more detailed information concerning the use of a grating in a particular configuration could be calculated simply and without further recourse to the computer. Their approach is somewhat abstract in that at some stages in the argument there is no obvious physical significance of the expressions which are derived. However, they start from a well known physical phenomenon and they finish by predicting on the basis of their theory further phenomena which were subsequently verified experimentally.

Consider the reflection of unpolarized light from the surface of a dielectric. It is well known that at angles of incidence other than normal the reflected light is to some extent polarized and that at one particular angle, the Brewster angle, this light is completely polarized. This occurs when, for light polarized with its electric vector in the plane of incidence, the electric vector of the refracted beam lies along the direction of the would-be reflected beam, as shown in Figure 6.17.

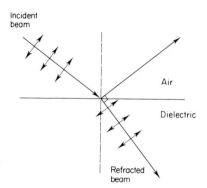

Figure 6.17 The Brewster condition for reflection.

Since the propagation of electromagnetic radiation requires a component of an electric field oscillating perpendicular to the direction of propagation, it follows that at the Brewster angle only light polarized with its electric vector perpendicular to the plane of incidence will be reflected. It follows from Snell's law that the Brewster angle is given by

$$\tan \theta_B = \mu \tag{6.9}$$

where μ is the refractive index of the dielectric, or that

$$\sin \theta_B = \frac{\mu}{(1 + \mu^2)^{1/2}} \tag{6.10}$$

What happens if we now replace the dielectric surface by a metal surface? A metal is characterized by a complex refractive index $n + ik$ where $i = \sqrt{-1}$, the real part n corresponds to the usual refractive index and k refers to the absorption. If we now insert this into expression (6.10), we obtain

$$\sin \theta_B = \alpha_z = \frac{n + ik}{(1 + n^2 - k^2)^{1/2}}$$

so that the sine of the Brewster angle is complex. We can attach no physical significance to this expression unless both α_z is real and $|\alpha_z|$ is less than unity. However, if there is a modulation on the surface of the metal the value of α_z changes. Maystre and Petit have calculated the variation of α_z and as the depth of modulation increases so α_z migrates across the complex plane as shown in Figure 6.18. What is more, it crosses the real axis so there exists for any given shape of groove (the ones shown here are sinusoidal) a value of the depth for which α_z has a real value of which the modulus is less than unity. In this case α_z has the physical significance of the sine of the Brewster angle for a modulated metal surface. In other words, the angle of incidence for which the light reflected from a grating is plane polarized. Maystre and Nevière (1977) showed that in the case of a grating which was so fine that only the zero order was allowed, the efficiency in that order could be calculated from the following simple formula:

$$E = R(\alpha_0) \left| \frac{\alpha_0 - \alpha_z(h)}{\alpha - \alpha_p(h)} \right|^2 \tag{6.11}$$

where α_0 is the sine of the angle of incidence, $R(\alpha_0)$ is the Fresnel reflection coefficient for the material of the grating, $\alpha_p(h)$ is a further complex number which represents the propagation constant of a surface wave (such as a plasmon) on the grating. $\alpha_p(h)$ represents a pole of E in the complex plane, and $\alpha_z(h)$ represents a zero. Their values as a function of groove depth h are shown in Figure 6.18 for sinusoidal gratings of pitch 1800 grooves mm^{-1}, covered with gold and used at a wavelength of 647 nm. From the figure we note two interesting features. Firstly, $\alpha_z(h)$ and $\alpha_p(h)$ tend to the same value as $h \rightarrow 0$. Therefore the expression reduces to the coefficient of Fresnel reflection for a plane surface. Secondly, $\alpha_z(h)$ crosses the real axis at a value of -0.115 when the depth of modulation is 40 nm. This means that when $\alpha_0 = \alpha_z(40)$ or $\sin^{-1}\theta = \sin^{-1}(-0.115) = -6.5°$, $E = 0$ and no light is reflected. Since there are no other orders for the light to go to, it follows that under these circumstances all

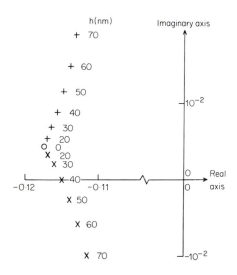

Figure 6.18 $\alpha_z(h)$ and $\alpha_p(h)$ plotted on an Argand diagram for various values of h: $+ = \alpha_p(h)$; $\times = \alpha_z(h)$.

the incident light is absorbed by the grating. We have the surprising result that a very gentle undulation in the surface is capable of transforming a highly reflecting gold mirror into one which absorbs all the light of one polarization incident upon it. This prediction has been verified experimentally and not only has the reduction of reflectance been observed, but it has also been possible to measure the rise in temperature of the grating due to the absorption of incident energy (Hutley and Maystre 1976). Figure 6.19 shows the comparison of experimental reflectance measurements and those predicted by Equation (6.11). Although the Brewster angle approach appears to be totally different from that of the surface plasmon explanation of anomalies, we must bear in mind that both models describe the same phenomenon. The sharp reduction of reflectance shown in Figure 6.19 can be regarded as an anomaly in the zero order and is no different in character from those already observed with gratings where more orders are able to propagate. Indeed, the approximate position of the anomaly can be calculated from Equation (6.7), which takes no account of the grating depth. At higher angles of incidence, true diffracted orders are allowed and if the curves in Figure 6.19 were carried on into this region they would strongly resemble those of Figure 6.11.

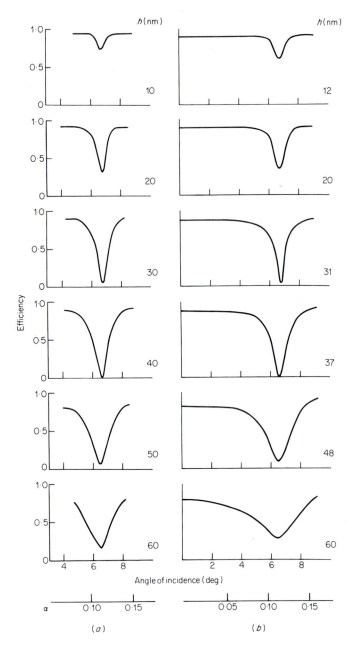

Figure 6.19 Theoretical (*a*) and experimental (*b*) reflectance curves for gold-coated, 1800 grooves mm^{-1} gratings of various depths.

The theory has been extended by Maystre, Nevière and Vincent (1978) to accommodate the case where several diffracted orders are allowed. Here the performance of a grating is characterized by a scattering matrix, the terms of which are simple functions of the zero and various poles of the complex plane. The significance of this work is that an otherwise enormous amount of data has been reduced to a manageable form and that it predicted the phenomenon of total absorption before it was observed. Thus the electromagnetic theory of gratings has passed from the stage of explaining observed results to predicting new ones and is therefore of real value to the optical designer because he can now test the performance of a grating without first having to make it.

☐   The effect of a dielectric overcoating on the anomalies

So far we have discussed only anomalies on gratings with a pure metal surface. In practice this is very seldom achieved and it is most usual to find that there is on top of the metal a thin layer of dielectric of one form or another. Most metals tarnish in the atmosphere, aluminium very soon forms a layer of oxide a few nanometres thick, silver forms a sulphide layer and only gold is completely inert. In some cases a layer of dielectric is applied intentionally in order to protect the surface either chemically or mechanically. For example, a thin layer of magnesium fluoride coated over aluminium will prevent oxidation and enable the component to be used in the vacuum ultraviolet down to a wavelength of 110 nm. On the other hand, the surface layer may arise accidentally by, for example, the deposition of a layer of vacuum pump oil on a grating used in the vacuum ultraviolet. What effect then does this have on the anomalies?

There are two possible effects and they depend upon the thickness of the dielectric layer. For a fairly thin layer the shapes and positions of the anomalies that are present on a bare metal grating are perturbed slightly. From the point of view of a grating used for spectroscopy in an instrument, the effect may be neglected and it is therefore possible to coat the gratings (and other optical components) with a thin protective layer in order to extend the life of commercial instruments. It is possible by studying the change in anomalies to determine the nature of the surface layer and Cowan and Arakawa (1970) have used the perturbations of the surface plasmons as a means of studying the grating surface.

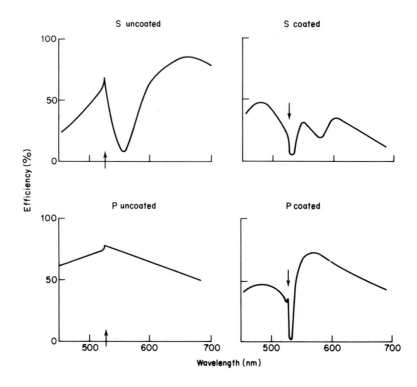

Figure 6.20 The effect of a dielectric layer on the anomalies of a grating.

If the dielectric layer is thicker than about 100 nm, then extra anomalies are seen to occur (Palmer 1952, Hutley *et al.* 1974). A typical example of this effect is shown in Figure 6.20, which shows the Littrow efficiency curves for a bare aluminium grating and one coated with 100 nm of silicon monoxide. In this case the S polarization surface plasmon anomaly splits into two and the Rayleigh anomaly changes from a cusp to a discontinuous shoulder. For P polarization there is a very strong narrow absorption very close to the Rayleigh wavelength. The introduction of a P polarization anomaly is consistent with the explanation of the redistribution of the energy in an order that is passing off. Since the surface is now dielectric, there is no reason why there should not be an electric field parallel to the surface. So the grazing order can carry a significant amount of energy and this will give rise to a significant perturbation in the remaining orders as it disappears.

A rather different though compatible explanation is that when the dielectric reaches a certain thickness it is capable of sustaining propagating surface modes just like those in a thin-film waveguide. Indeed one of the ways of coupling energy into and out of guided waves in thin films is to incorporate a grating at each end of the guide. This in itself is a subject which is attracting a great deal of attention because of the recent increase in interest in optical guided waves.

☐   The repulsion of anomalies

If one calculates on the basis of simple theory the positions at which anomalies are expected to occur, it is apparent that there are various positions on the efficiency surface where two anomalies occur together. In the case of the Rayleigh anomalies these points lie at the intersections of the curves describing the passing off of the various orders (Figure 4.16). The same is true if we adopt Equation (6.7) for surface plasmon anomalies: we obtain a similar set of curves, but they are displaced slightly depending upon the grating material. The simultaneous occurrence of Rayleigh anomalies corresponds to the passing off of a positive order in one direction and a negative order in the other, as indicated in Figure 6.21. In the case of surface plasmon anomalies it corresponds to the simultaneous creation of surface plasmons in opposite directions.

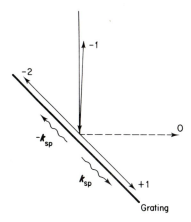

Figure 6.21 The simultaneous occurrence of Rayleigh anomalies.

In practice it was found that the anomalies did not merge at these points but remained separate. In 1962 Stewart and Galloway plotted the exact position of the anomalies in the region of a crossover and obtained results similar to those shown in Figure 6.22. The experimental points closely followed the loci of the calculated Rayleigh wavelengths except where the two anomalies should coincide. Instead of crossing, the anomalies appeared to repel each other, exchange identitites and then separate. At that time the distinction between Rayleigh anomalies and surface plasmon or resonance anomalies was not appreciated, but it now appears that it is the latter that are observed to repel each other. From the point of view of quantum mechanics, the repulsion of anomalies is seen as a discontinuity in the dispersion curve of the surface plasmons. This was explained by Cowan and Arakawa (1970) in terms of the interaction of surface plasmons having the same energy but propagating in different directions and forming standing waves in the periodic electron density of the grating surface. The phenomenon is analogous to gaps in the Brillouin zones of crystalline solids.

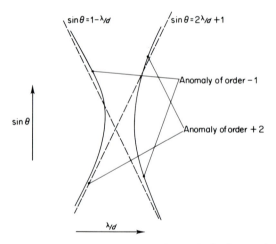

Figure 6.22 The observed (———) and theoretical (———) respulsion of anomalies. (After Stewart and Galloway 1962.)

It does not always happen that the anomalies repel; on some occasions they merge and, as it were, pass through each other, each unaffected by the other. In Figure 6.23 we show some measurements of the anomalies of two interference gratings. In one case the anomalies merge and in the other they repel. What is perhaps most surprising

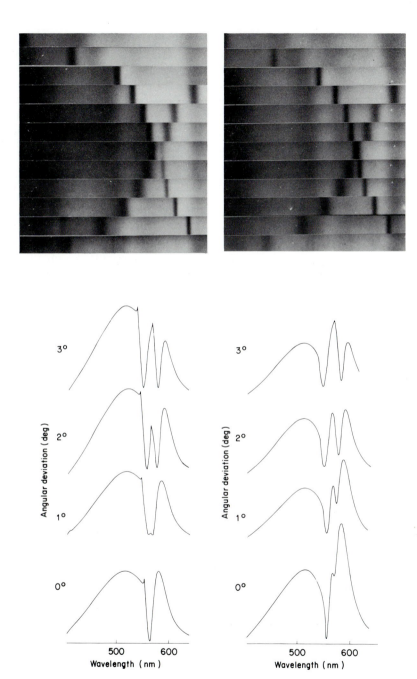

Figure 6.23 Examples of anomalies: (*a*) merging and (*b*) repelling.

about these results is that there is so little difference between the measured shapes of the groove profiles. It is also interesting to note that the character of the Rayleigh anomalies is quite different in the two cases and it is not clear why this should be so.

☐   Summary

It has been established that the theory of diffraction grating efficiency has reached the stage where it can adequately describe the performance of a grating provided that sufficient data are available to describe the shape and nature of the grooves. In general it is important to take into account the optical properties of the grating surface, but under some circumstances it is possible to use a simplified theory based on the assumption that the grating has infinite conductivity at the surface. This will be true for wavelengths greater than about $4\,\mu m$ where most metals can be considered effectively as perfect conductors. The ability to predict with confidence the performance of a grating before it is even made may now be used in the design of gratings and this may save a significant amount of time and effort, particularly if the gratings have to be ruled. As we have seen, it has already predicted some rather surprising results.

It is one thing to be able to describe how a given grating behaves, but it is another to understand how gratings work. The description of anomalies in terms of surface plasmons and in terms of Brewster angle effects has given us some considerable insight into this but, at the time of writing, there still remain some gaps in our understanding. We still cannot pinpoint which features of a grating (symmetry of profile, straightness of facets, sharpness of corners, roughness of surface, etc.) are likely to induce anomalies. Nor can we say without recourse to the computer, whether an anomaly will occur as a bright or a dark band, or as a cusp or a shoulder. Furthermore, there remains the inverse diffraction problem: that is, given the efficiency, deduce the groove profile. A solution to this problem may well tell us whether the types of grating that are used at present are the most efficient, or whether there exists an entirely different shape of groove that would give higher efficiencies over a greater range of wavelengths. A great deal of progress has been made in this direction by Roger and Maystre (1979) at Marseille and in some cases the problem has been solved.

For the spectroscopist these questions tend to be a little academic. His main concern is simply that anomalies are likely to exist and he is well advised to measure them as part of the procedure of calibrating his instrument. This involves measuring the spectrum from a polarized continuous source. In the visible a tungsten filament and polarizing filter will suffice, though difficulty may be experienced at shorter wavelengths. It is also an advantage to be able to calculate approximately where in the spectrum anomalies are likely to occur. It may very well be that discrepancies between spectra recorded on different instruments are due to anomalies and this possibility should often be explored before searching for deeper explanations. Just as a ghost can masquerade as a line in an emission spectrum so may an anomaly be mistaken for an absorption band.

Appendix 6.1

☐ The reciprocity theorem in the case of a
grating where the grooves are vignetted

According to the reciprocity theorem, the flux passing through an optical system is independent of the direction of travel through the system. At first sight this seems surprising in the case of a grating, particularly an echelle, where the angles of incidence and diffraction are different, as shown in Figure 6.24.

In the first case, Figure 6.24(a), the angle of incidence α is less than the angle of diffraction β and a proportion of the incident beam is vignetted. Either it strikes the wrong facet on the way in or it is obstructed on the way out and the useful fraction of the beam is $\cos B/\cos A$. On the other hand, when the direction of radiation is reversed, vignetting no longer occurs and all the beam is used (Burton and Reay 1970, Schroeder and Hilliard 1980). In one case energy is absorbed or scattered by the grating and in the other it is not. Does this not violate the reciprocity theorem?

Let us consider the spectral image as shown in Figure 2.25. It consists of a Dirac comb (the orders of diffraction) convoluted with a sinc function (due to the finite width of the grating) and multiplied by an envelope function (which describes the blaze effect of the grooves). In the case of an echelle we may use simple scalar theory and regard the envelope function as the diffraction pattern of each groove considered as a slit.

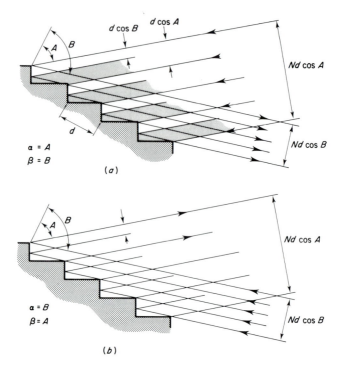

Figure 6.24 The vignetting of a grating depending upon the direction of incidence.

The total flux in a given diffracted order is represented by the area under the curve relating to that order, and is proportional to the product of the peak amplitude and the width. The width is, of course, inversely proportional to the physical width of the diffracted beam. In the unvignetted case, $\alpha < \beta$ (Figure 6.24(b)), the width of the beam is $Nd \cos A$ and in the vignetted case, $\alpha > \beta$ (Figure 6.24(a)), it is $Nd \cos B$. The widths of the two spectral images are therefore in the ratio

$$\frac{\Delta\beta(\alpha < \beta)}{\Delta\beta(\alpha > \beta)} = \frac{\cos B}{\cos A}$$

but their amplitudes are in the ratio

$$\frac{A_0(\alpha < \beta)}{A_0(\alpha > \beta)} = \frac{\cos A}{\cos B}$$

The two effects cancel and the flux is independent of the direction of the light through the system (Bottema 1980).

It is true that the peak irradiance is not constant and that the total flux integrated over all orders is not constant, but the reciprocity theorem still holds.

Bibliography

For a review of the electromagnetic theory of diffraction gratings, see:

R. Petit (1980). "The Electromagnetic Theory of Gratings". Spinger, Berlin.

and, in addition to the references quoted in the text:

Petit and Maystre (1972)[†]
Kalhor and Neureuther (1971)[†]
Van den Berg (1971)[†]
Jovecevic and Sesnic (1976)[†]

For further reading on surface plasmons:

Ritchie (1973)[†]
Fischer *et al.* (1973)[†]
Pockrand (1976)[†]

[†] See main reference list.

7

Concave Gratings

So far we have considered only gratings that are plane and have straight, equally spaced grooves, so that the same terms in the grating equation apply over the whole of the grating area. In order that the angle of incidence be constant, the incident light must be collimated, which is usually achieved with a lens or mirror. The groove spacing is constant, so the angle of diffraction is constant and a further component is required to bring the diffracted light to a focus. In 1882 H. A. Rowland showed that by forming a grating on a spherical concave substrate, the grating itself could perform the task of both dispersing the light and bringing it to a focus. This is of particular advantage if a significant proportion of the light is lost at the collimating and focusing components. In Rowland's time most reflecting optical components were made of speculum metal which has at best a reflectance R of only 70%. The maximum transmission through an instrument with a plane grating (with a supposed relative efficiency of 100%), would then be R^3 or 35%, which is just half of that of an instrument with a single concave grating. For visible light this advantage was eroded when vacuum-deposited thin films of metals such as aluminium and silver became available. With a typical reflectance of 90%, the transmittances of the two types of instrument would then be 90% and 73%. However, at shorter wavelengths the reflectance of these metals falls dramatically as we see from Figure 7.1. For wavelengths down to 200 nm, bare aluminium is suitable and its use may be extended down to 110 nm by overcoating with a protective layer of magnesium fluoride. However, below this wavelength the best one can hope to

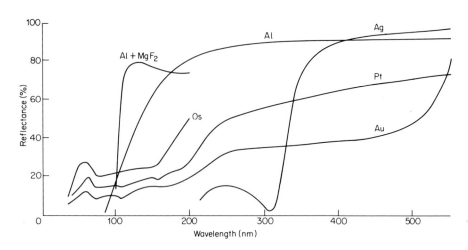

Figure 7.1 The reflectance of metals in the ultraviolet.

achieve is a reflectance at normal incidence of about 20%. The transmission through an instrument with a concave grating is then only 20%, and that through a plane grating instrument is a mere 0.8%. For spectroscopy in this region a concave grating is therefore essential.

Rowland ruled gratings on concave spherical substrates using the same ruling engine that was used for ruling plane gratings. Under these circumstances, illustrated in Figure 7.2, the spacing is not constant around the grating surface, but it is constant along a chord. It also happens that the facet angle of a blazed grating remains constant with respect to that chord and not to the grating surface.

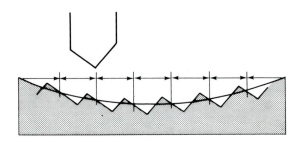

Figure 7.2 A conventional concave ruled diffraction grating.

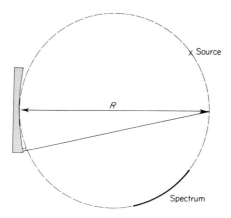

Figure 7.3 The Rowland circle.

Rowland showed that for a grating made in this way, if a point source is located on a circle which touches the pole of the grating, but which has half the radius of curvature of the blank, then to a first approximation the diffracted image will also be on that circle. This circle is known as the Rowland circle and is shown in Figure 7.3. We shall first give a simplified explanation of the Rowland circle and then show that it is a particular solution to a more general expression governing the focal properties of spherical concave gratings. We shall then show how the general expression may be generalized still further to take account of gratings with grooves that are not straight and equally spaced but are the loci of interference fringes recorded "holographically".

□   The Rowland circle

We shall first consider a small area of grating around the pole of the sphere. We restrict ourselves to a single plane perpendicular to the rulings so that we neglect the effect of the length of the ruling, and we also neglect the fact that the grating does not coincide with the Rowland circle but is a tangent to it.

In Figure 7.4, C is the centre of curvature of the grating, a source of light of wavelength λ is situated at P and is brought to a focus at Q. The conditions of diffraction at A are governed by the grating equation:

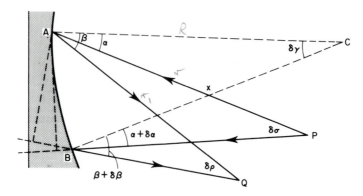

Figure 7.4 Parameters for the calculation of the Rowland circle condition.

$$d(\sin \alpha + \sin \beta) = m\lambda$$

which we may differentiate to obtain

$$\cos \alpha \, \partial\alpha + \cos \beta \, \partial\beta = 0 \qquad (7.1)$$

If we now consider the triangles ACX and PBX we see that

$$\alpha + \delta\gamma = \alpha + \delta\alpha + \delta\sigma$$

so that

$$\delta\alpha = \delta\gamma - \delta\sigma$$

and similarly

$$\delta\beta = \delta\gamma - \delta\rho \qquad (7.2)$$

However, from the figure we see that since $\delta\alpha$, $\delta\beta$, $\delta\gamma$, $\delta\sigma$ are small

$$\delta\gamma = \frac{AB}{R}, \qquad \delta\sigma = \frac{AB \cos \alpha}{r}, \qquad \delta\rho = \frac{AB \cos \beta}{r_1}$$

where $R = AC$ = radius of the grating, $AP = r$ and $AQ = r_1$. So we may substitute for $\partial\alpha$ and $\partial\beta$ in Equation (7.1) to obtain

$$\frac{\cos \alpha}{R} - \frac{\cos^2\alpha}{r} + \frac{\cos \beta}{R} - \frac{\cos^2\beta}{r_1} = 0 \qquad (7.3)$$

which is the focal equation for the spherical concave grating.

One solution of this equation is that the terms in α and β vanish separately, in which case,

$$\frac{\cos \alpha}{R} = \frac{\cos^2\alpha}{r} \quad \text{and} \quad \frac{\cos \beta}{R} = \frac{\cos^2\beta}{r_1} \qquad (7.4)$$

so

$$r = R \cos \alpha \quad \text{and} \quad r_1 = R \cos \beta$$

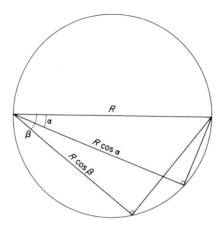

Figure 7.5 The construction of the Rowland circle.

which we see from Figure 7.5 is the condition that both P and Q lie on a circle of *diameter R*.

Let us now consider the requirements for the groove spacing at different points on the grating surface for the Rowland circle condition. For simplicity of the geometry and without loss of generality we consider the formation of the image of a source situated at the centre of curvature of the gratings, as shown in Figure 7.6.

Consider diffraction from two grooves VV′ at the pole of the grating and QQ′ at a different point on the grating surface. Since the source is at the centre of curvature of the blank, at both grooves the light arrives at normal incidence. The condition that the diffracted light be brought to a focus at P is then that

$$m\lambda = d \sin \beta$$

and

$$m\lambda = d' \sin \beta'$$

Therefore

$$d' = \frac{d \sin \beta}{\sin \beta'} \tag{7.5}$$

If we now apply the sine rule to triangle PQ′N we see that

$$\frac{\sin \beta'}{PN} = \frac{\sin (90 - \theta + \Delta)}{Q'N} = \frac{\Delta \sin \theta + \cos \theta}{R}$$

where $\Delta = \beta - \beta'$ and is small so that we may write $\cos \Delta = 1$ and $\sin \Delta = \Delta$. Now $PN = R \sin \beta$ so

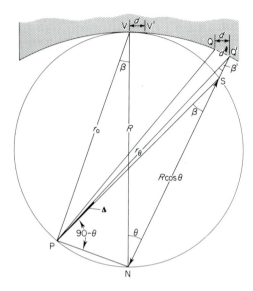

Figure 7.6 The variation of pitch across the aperture required for Rowland circle focusing.

$$\frac{\sin \beta}{\sin \beta'} = \frac{1}{\Delta \sin \theta + \cos \theta} = \frac{1}{\cos \theta}$$

if $\Delta \sin \theta$ is sufficiently small to be neglected. We may then write

$$\frac{\sin \beta}{\sin \beta'} = \frac{1}{\cos \theta} \quad \text{and} \quad d' = \frac{d}{\cos \theta}$$

That is to say, the Rowland circle focal condition is met when the pitch of the grating varies in such a way that its projection on a chord is constant. It is, of course, just such a variation that is generated when a grating is ruled on a conventional ruling engine.

☐   **Spherical aberration**

If we consider gratings of larger aperture, θ is not very small and the term $\Delta \sin \theta$ can no longer be neglected. Grooves ruled according to a $1/\cos \theta$ law are in fact slightly too wide, so the angle of diffraction is slightly too small and rays diffracted from the edge of the grating do not come to a focus exactly on the Rowland circle. The image suffers from spherical aberration. The error Δp in path difference across a groove is given by

$$\Delta p = d \sin \beta \left[\frac{1}{\Delta \sin \theta + \cos \theta} - \frac{1}{\cos \theta} \right] \tag{7.7}$$

and it may be shown (Madden and Strong 1958) that the total error in path length between rays diffracted from the pole of the grating and from a region subtending an angle θ is equal to

$$\frac{m\lambda R \tan \beta' \theta^4}{8d} \tag{7.8}$$

The important feature to appreciate is that the error increases with the fourth power of the numerical aperture, so that when designing an instrument it is often necessary to compromise between the light-gathering power, which increases with the square of the numerical aperture, and the resolution which, through aberration, decreases with the fourth power of the numerical aperture.

☐   Astigmatism

The most dramatic aberration of the spherical concave grating when used in a Rowland circle mounting is that of astigmatism. Here a point source is focused not as a point, but as a line. Let us consider the formation of an image of a point on the Rowland circle, and let us assume for the sake of simplicity that in the plane of dispersion (the meridional plane) the distances from the source and the image to the grating are equal (Figure 7.7). The image/object distance is given by $R \cos \theta$ and since the grating is imaging a point back into itself it is behaving as if it were a concave mirror of radius $R \cos \theta$. However, in the vertical plane the rays are on-axis and the effective radius of curvature is unaltered and equal to R.

  Thus the image in one plane is formed at a different position from that formed in the other and the image formed in the vertical plane will be out of focus at the position of the image which is formed in the plane of dispersion. At this position the image of a point source will take the form of a line of finite length, as indicated in Figure 7.7.

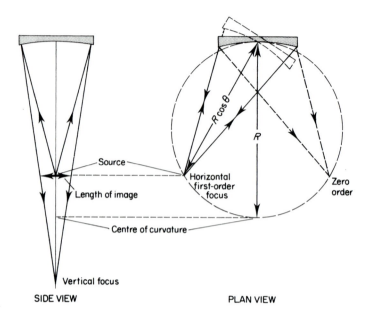

Figure 7.7 The source of astigmatism in concave gratings.

☐   A generalized treatment of the focal properties
            of a concave grating

The discussion so far has involved many simplifying assumptions
and, although it is valid within the terms of those assumptions,
it is cumbersome to extend to the more rigorous case. The main
advantage is that it has enabled us to bring out certain features
of the concave grating, namely the focal equation of which the
Rowland circle is an example, the fact that the groove density
projected onto a chord should be constant, the astigmatic nature of
the focus and the existence of spherical aberration. A far more
elegant treatment, and one which lends itself to greater rigour,
involves the application of Fermat's principle to the optical path
length from the source to the image via any point on the grating
(Welford 1965).

Consider light from a point A in space passing via a point P on the
grating to form a focus at a point B. We can describe the geometry
with respect to a set of cartesian coordinates with the origin at the
pole of the grating (Figure 7.8).

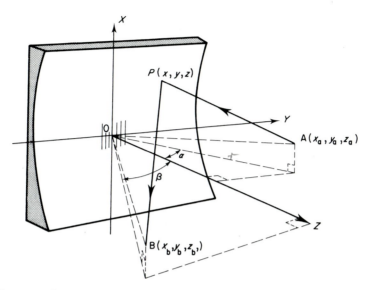

Figure 7.8 Parameters for the construction of the general focal condition.

The optical path W is $\langle AP \rangle + \langle PB \rangle$ and the conditions that B should be the position of the focused image of A are, first, that as P moves along the groove the path difference should remain constant and, second, that the change in path in going from one groove to the next is $m\lambda$ so that

$$\frac{\partial W}{\partial x} = 0, \qquad \frac{\partial W}{\partial y} = \frac{m\lambda}{d} \qquad (7.9)$$

(That this is constant implies that the projection of the groove spacing on the Y axis is constant.)

The path length may be calculated from the coordinates of A, P and B such that

$$\langle AP \rangle = [(x_a - x)^2 + (y_a - y)^2 + (z_a - z)^2]^{1/2}$$

$$\langle PB \rangle = [(x_b - x)^2 + (y_b - y)^2 + (z_b - z)^2]^{1/2} \qquad (7.10)$$

which can be expanded in a Taylor series.

This expression is valid for any shape of grating described by a function

$$z = f(x, y)$$

For a sphere of radius R we have

$$(z - R)^2 + x^2 + y^2 = R^2$$

so that we may substitute for z in Equation (7.10) (although this in turn requires a series expansion). It is also convenient to express the position of A and B in cylindrical coordinates, in which

$$x_a = r \cos \alpha, \qquad x_b = r_1 \cos \beta$$
$$y_a = r \sin \alpha, \qquad y_b = r_1 \sin \beta$$

The resulting expression is a power series in x and y of which the first few terms are:

$$W = r\left(1 + \frac{x_a^2}{r^2}\right)^{1/2} + r_1\left(1 + \frac{x_b^2}{r_1^2}\right)^{1/2}$$

$$- y\left[\left(1 + \frac{x_a^2}{r^2}\right)^{-1/2} \sin \alpha + \left(1 + \frac{x_b^2}{r_1^2}\right)^{-1/2} \sin \beta\right]$$

$$- x\left[\frac{x_a}{r}\left(1 + \frac{x_a^2}{r^2}\right)^{-1/2} + \frac{x_b}{r_1}\left(1 + \frac{x_b^2}{r_1^2}\right)^{-1/2}\right]$$

$$+ \tfrac{1}{2}y^2\left[\left(1 + \frac{x_a^2}{r^2}\right)^{-3/2}\left(\frac{\cos^2\alpha}{r} - \frac{\cos \alpha}{R} + \frac{x_a^2}{r^2}\left\{\frac{1}{r} - \frac{\cos \alpha}{R}\right\}\right)\right.$$

$$+ \left.\left(1 + \frac{x_b^2}{r_1^2}\right)^{-3/2}\left(\frac{\cos^2\beta}{r'} - \frac{\cos \beta}{R} + \frac{x_b^2}{r_1^2}\left\{\frac{1}{r'} - \frac{\cos \beta}{R}\right\}\right)\right]$$

$$+ \tfrac{1}{2}x^2\left[\left(\frac{1}{r} - \frac{\cos \alpha}{R}\left\{1 + \frac{x_a^2}{r^2}\right\}\right)\left(1 + \frac{x_a^2}{r^2}\right)^{-3/2}\right.$$

$$+ \left.\left(\frac{1}{r'} - \frac{\cos \beta}{R}\left\{1 + \frac{x_b^2}{r_1^2}\right\}\right)\left(1 + \frac{x_b^2}{r_1^2}\right)^{-3/2}\right]$$

$$- xy[\,\ldots\,]\ldots$$

$$= W_{00} + x W_{10} + y W_{01} + \tfrac{1}{2}x^2 W_{20} + \tfrac{1}{2}y^2 W_{02} + xy W_{11} + \ldots$$

$$(7.11)$$

Although this expression is somewhat cumbersome, it is convenient to consider the various terms individually because each has a different physical significance. The first-order terms yield the grating equation, the second-order terms the condition for the position of an image, and the higher-order terms describe astigmatism, coma, spectral line curvature, spherical aberration and so on.

If we first apply the conditions

$$\frac{\partial W}{\partial x} = 0 \quad \text{and} \quad \frac{\partial W}{\partial y} = \frac{m\lambda}{d}$$

to the principal ray, that is the ray which passes via the pole of the grating where $x = y = z = 0$, we obtain the following expression:

$$W_{10} = 0 = \frac{x_a}{r}\left(1 + \frac{x_a^2}{r^2}\right)^{-1/2} + \frac{x_b}{r_1}\left(1 + \frac{x_b^2}{r_1^2}\right)^{-1/2} \tag{7.12a}$$

and

$$W_{01} = \left(1 + \frac{x_a^2}{r^2}\right)^{-1/2}\sin\alpha + \left(1 + \frac{x_b^2}{r_1^2}\right)^{-1/2}\sin\beta = \frac{m\lambda}{d} \tag{7.12b}$$

The first equation has a solution

$$x_a/r = -x_b/r'$$

which in effect says that in a vertical plane the grating acts as a mirror and the angle of incidence in the vertical plane $(\tan^{-1}(x_a/r))$ is equal to the angle of reflection $(\tan^{-1}(x_b/r_1))$.

If we substitute this result in Equation (7.12b) we have:

$$\left(1 + \frac{x_a}{r}\right)^{1/2}\{\sin\alpha + \sin\beta\} = \frac{m\lambda}{d} \tag{7.13}$$

which is the grating equation modified to take account of incidence out of a plane which is normal to both the grating and the grooves. The condition for a vertical focus, in effect when the grating has a finite width but grooves of negligible length, is obtained by setting the derivative of the second-order term W_{02} equal to zero which yields

$$\left(1 + \frac{x_a^2}{r^2}\right)^{-3/2}\left\{\frac{\cos^2\alpha}{r} - \frac{\cos\alpha}{R} + \frac{x_a^2}{r_1^2}\left[\frac{1}{r} - \frac{\cos\alpha}{R}\right]\right\}$$
$$+ \left(1 + \frac{x_b^2}{r_1^2}\right)^{-3/2}\left\{\frac{\cos^2\beta}{r_1} - \frac{\cos\beta}{R} + \frac{x^2}{r_1^2}\left[\frac{1}{r_1} - \frac{\cos\beta}{R}\right]\right\} = 0 \tag{7.14}$$

If at this stage we were to set $x_a = x_b = 0$, so that the problem reduces to that where all points lie in the YZ plane, we see that

$$\frac{\cos^2\alpha}{r} - \frac{\cos\alpha}{R} + \frac{\cos^2\beta}{r_1} - \frac{\cos\beta}{R} = 0$$

which is the focal equation we derived earlier (Equation (7.3)).

The detailed analysis of the aberrations of concave gratings arising from the higher-order terms in Equation (7.11) has been described in full by several authors, in particular Beutler (1945), Namioka (1959) and Danielsson and Lindblom (1975), and we shall now summarize some of their results. In most cases however, their analysis is slightly different in that they apply Fermat's principle to the "light path function" $F = \langle AP \rangle + \langle PB \rangle + y$ rather than to the optical path $W = \langle AP \rangle + \langle PB \rangle$. The light path function represents the difference in optical path between a ray passing through a general point and a ray passing through the origin. In this case the application of Fermat's principle takes the form

$$\frac{\partial F}{\partial y} = \frac{\partial F}{\partial x} = 0 \qquad (7.15)$$

rather than that of Equation (7.9). Both lead to the same result. The first-order terms in the power series Equation (7.11) lead to the grating equation in y and the laws of reflection in x. The terms in x^2 and y^2 lead respectively to the horizontal and vertical focusing conditions. Since we choose the vertical focus in a spectrometer, the x term describes the astigmatism. The cross-terms describe spectral line curvature, the third-order terms coma and fourth-order terms spherical aberration. In the case of the Rowland circle mountings the terms W_{02}, W_{03}, W_{04} vanish and in particular there is no coma. The length of the astigmatic focal line is given by

$$L = H \cos \beta \left(\frac{\sin^2 \alpha}{\cos \alpha} + \frac{\sin^2 \beta}{\cos \beta} \right) \qquad (7.16)$$

where H is the height of the ruling. For a point source this astigmatism does not affect the width of the spectral image and hence the resolution of an instrument, but it does give rise to a serious loss of irradiance since the incident flux is spread over a far greater area than for a good focus. In practice, however, it is usually necessary to ensure that the incident radiation illuminates an entrance slit which, allowing for magnification, is effectively the same length as the exit slit. Unfortunately, however, the spectral image for a point source is curved because rays which are incident upon the grating out of the meridional plane YZ are deviated by a different amount from those within the meridional plane. The spectral images are concave towards longer wavelengths and their curvature projected in a plane perpendicular to the principal ray is given by

$$C = \frac{\sin \alpha + \sin \beta}{r' \cos \beta} \qquad (7.17)$$

When the entrance slit has a finite length, each point on that slit produces its own curved image and these lie on another curve which may have the same or the opposite sign. The result is that the two curves are convoluted (Figure 7.9) and this does give rise to a loss of resolution.

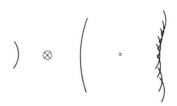

Figure 7.9 The effect of image curvature in the instrumental linewidth.

Under Rowland circle conditions, the spherical aberration expressed as the difference in image distance r_1 for rays passing via the origin from those passing via a point (θ, y) on the grating meridian is

$$\frac{y^4}{8R^2} \left[\frac{\cos^2 \alpha}{\sin \alpha} + \frac{\cos^2 \beta}{\sin \beta} \right] \qquad (7.18)$$

(cf. Equation (7.8) which describes the change in phase of the wavefront due to spherical aberration.)

As we have already remarked, this expression increases very rapidly with the width of the grating and in order to maintain a compromise between resolving power and flux through the instrument it is necessary to limit the width of the grating. The optimum value will in fact depend not only upon the resolution that is required but also upon the length of the grooves since this also affects the flux and, through astigmatism and line curvature, the resolution. Mack, Stehn and Edlen (1932) calculated that in the meridional plane the maximum width of grating compatible with the achievement of maximum theoretical resolution was given by:

$$W_{\text{opt}} = 2.36 \left[\frac{4\lambda R^3}{\pi(\cot \alpha \cos \alpha + \cos \beta \cot \beta)} \right]^{1/3} \qquad (7.19)$$

□   Mountings of the concave grating

We have derived a general equation which describes the relationship between the position of a point source and that of a spectral image in the meridional plane of a spherical concave grating. This takes the form:

$$\frac{\cos^2\alpha}{r} - \frac{\cos\alpha}{R} + \frac{\cos^2\beta}{r_1} - \frac{\cos\beta}{R} = 0$$

(Equation (7.3)) and refers to a grating of radius R in which the projected groove spacing is constant along a chord of the grating blank. The first well known solution to this equation corresponds to the Rowland circle mounting in which the terms in α and in β are set separately to zero so that

$$R\cos\theta = r \quad \text{and} \quad R\cos\beta = r_1$$

There are various ways in which the Rowland circle condition may be satisfied and some of them are shown in Figure 7.10. The simplest mounting of all is that due to Paschen and Runge (Figure 7.10(a)) in which the entrance slit is positioned on the Rowland circle and a photographic plate (or plates) is constrained to fit the Rowland circle. Alternatively, for photoelectric detection, a series of exit slits is arranged around the Rowland circle each having its own detector. In the latter case the whole spectrum is not recorded, only a series of predetermined wavelengths, but when used in this "polychromator" form it is very rugged and convenient for applications such as the routine analysis of samples of metals and alloys. In this case slits and detectors are set up to measure the light in various spectral lines, each characteristic of a particular component or trace element.

The oldest concave grating mount is that designed by Rowland himself and which bears his name (Figure 7.10(b)). In this case the grating and photographic plates are fixed at opposite ends of the diameter of the Rowland circle by a moveable rigid beam. The entrance slit remains fixed above the intersection of two rails at right angles to each other and along which the grating plate holder (or the exit slit) are free to move. In this way this entrance slit, grating and plate holder are constrained always to lie on the Rowland circle and it has the advantage that the dispersion is linear, which is useful in the accurate determination of wavelengths. Unfortunately, this mounting is rather sensitive to small errors in the position of the entrance slit and in the orthogonality of the rails and is now very rarely used.

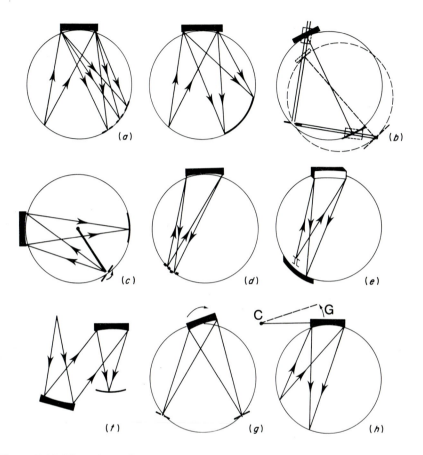

Figure 7.10 Mountings of the concave grating: (*a*) Paschen–Runge; (*b*) Rowland; (*c*) Abney; (*d*) and (*e*) Eagle; (*f*) Wadsworth; (*g*) Seya–Namioka; (*h*) Johnson–Onaka.

A variation on this mounting was devised by Abney who again mounted the grating and the plate holder on a rigid bar at opposite ends of the diameter of the Rowland circle (Figure 7.10(*c*). The entrance slit is mounted on a bar of length equal to the radius of the Rowland circle. In this way the slit always lay on the Rowland circle, but it had to rotate about its axis in order that the jaws should remain perpendicular to the line from the grating to the slit. It also had the disadvantage that the source must move with the exit slit, which could well be very inconvenient.

In the Eagle mounting (Eagle 1910) the angles of incidence and diffraction are made equal, or very nearly so, as with Littrow mounting for plane gratings. Optically this system has the advantage that the astigmatism is generally less than that of the Paschen–Runge or Rowland mounting, but on the other hand the dispersion is non-linear. From a mechanical point of view it has the disadvantage that it is necessary with great precision both to rotate the grating and to move it nearer to the slits in order to scan the spectrum. However it does have the great practical advantage that it is very much more compact than the other mountings, and this is of particular import-ance when we bear in mind the need to enclose the instrument in a vacuum tank. Ideally the entrance slit and exit slit or photographic plate should be superimposed if we are to set $\alpha = \beta$. In practice, of course, the two are displaced either sideways, in the plane of incidence as shown in Figure 7.10(d), or out of plane of incidence, in which case the entrance slit is positioned below the meridonal plane and the plate holder just above it, as shown in Figure 7.10(e). The out-of-plane configuration is usually used for spectrographs and the in-plane system for monochromators. The penalty incurred in going out of plane is that coma is introduced in the image and slit curvature becomes more important. This limits the length of the ruling that can effectively be used.

The second well known solution to Equation (7.3) is the Wadsworth mounting (Wadsworth 1903; Figure 7.10(f)) in which the incident light is collimated, so r is set at infinity and the focal equation reduces to

$$ r_1 = \frac{R \cos^2 \beta}{\cos \alpha + \cos \beta} \tag{7.20} $$

One feature of this mounting is that the astigmatism is zero when the image is formed at the centre of the grating blank, i.e. when $\beta = 0$. It is a particularly useful mounting for applications in which the incident light is naturally collimated (for example in rocket or satellite astronomy, spectroheliography and in work using synchrotron radiation). However, if the light is not naturally collimated the Wadsworth mount requires a collimating mirror, so one has to pay the penalty of the extra losses of light at this mirror. If the spectrum is to be recorded photoelectrically this loss may be very severe — a factor of five or ten perhaps — but if the spectrum is to be recorded photographically, then it may happen that the increase of irradiance resulting from the decrease in the length of the image can offset the extra loss of flux at the collimating mirror. The distance from the grating to the image is about half that for a Rowland circle mounting which makes the instrument more compact and, since the grating subtends approximately four times the solid angle, there is a corre-sponding increase in the brightness of the spectral image.

Not all concave grating mountings are solutions to Equation (7.3). In some cases other advantages may compensate for a certain defect of focus. A particularly important example of this is the Seya–Namioka mounting in which the entrance slit and exit slit are kept fixed and the spectrum is scanned by a simple rotation of the grating (Seya 1956; Figure 7.10(g)). In order to achieve the optimum conditions for this mounting we set $\alpha = \phi + \theta$ and $\beta = \theta - \phi$, where 2ϕ is the angle subtended at the grating by the entrance and exit slit and θ is the angle through which the grating is turned. The amount of defocus is given by

$$F(\theta, \phi, r, r_1) = \frac{\cos^2(\theta + \phi)}{r} + \frac{\cos(\theta + \phi)}{R} + \frac{\cos^2(\theta - \phi)}{r_1} + \frac{\cos(\theta - \phi)}{R}$$

$$(7.21)$$

and the optimum conditions are those for which $F(\theta, \phi, r, r_1)$ remains as small as possible as θ is varied over the required range. Seya set F and three derivatives of F with respect to θ equal to zero for $\theta = 0$ and obtained the result:

$$\phi = \sin^{-1}(1/\sqrt{3}) = 35°15'$$

and

$$r = r_1 = R\cos\phi$$

which corresponds to the Rowland circle either in zero order or for zero wavelength. In practice it is usual to modify the angle slightly so that the best focus is achieved in the centre of the range of interest rather than at zero wavelength.

The great advantage of this mounting is its simplicity. An instrument need consist only of a fixed entrance and exit slit and a simple rotation of the grating is all that is required to scan the spectrum. It is in fact simpler than instruments using plane gratings. Despite the fact that at the ends of the useful wavelength range the resolution is limited by the defect of focus and the astigmatism is particularly bad, the Seya–Namioka mounting is very widely used, particularly for medium-resolution rather than high-resolution work.

A similar simplicity is a feature of the Johnson–Onaka mounting (Johnson 1957). Here again the entrance and exit slits remain fixed, but the grating is rotated about an axis which is displaced from its centre; in this way it is possible to reduce the change of focus that occurs in the Seya–Namioka mounting. The system is set up so that at the centre of the desired wavelength range the slits and grating lie on the Rowland circle, as shown in Figure 7.10(h). The optimum radius of rotation, i.e. the distance GC, was found by Onaka (1958) to be

$$GC_{opt} = \frac{R \sin \frac{1}{2}(\alpha + \beta)}{1 - \frac{1}{2} \tan(\alpha + \beta)[\tan\beta - \tan\alpha]} \qquad (7.22)$$

☐   Aberration-corrected gratings

All that has been said so far has related to gratings on spherical substrates where the grooves are straight and equally spaced when projected on a chord. This is the type of grating that is obtained by substituting a concave spherical blank in place of a plane one on a conventional ruling engine. The aberrations that are present in a spectral image are a consequence of the form of the grating and can be varied and in some cases eliminated if we allow the grating to take a different form. There are three ways in which we can do this. We change the shape of the substrate or of the grooves and we may vary the spacing of the grooves across the grating. The possibility of controlling the focal properties of gratings by a proper distribution of the grooves has been known for a long time (Haber 1950, Gale 1966, Sakayanagi 1967). It was discussed, for example, by Cornu in some detail in 1893 when he predicted the effects of varying the groove spacing in both plane and concave gratings. However, because of the considerable technical difficulties of manufacturing even regular gratings, this work remained of only academic interest until comparatively recently. The possibility of making interference gratings has led to a resurgence of interest, because the technique enables gratings to be made on substrates that would be difficult to rule upon (e.g. too steep) and enables the groove spacing and shape to be varied by using non-planar wavefronts (Cordelle *et al.* 1969). The way in which these techniques may be applied to obtain a good focused image may conveniently be understood by applying the rules of holography. Indeed, under these circumstances the term "holographic grating" may be used rather more appropriately than when describing interference gratings in general.

Consider for example, an interference grating made on a concave spherical substrate as shown in Figure 7.11. The interfering wavefronts are derived from two point sources, one of which is situated at the centre of curvature of the blank and the other at some arbitrary position.

We may regard the first as an object beam, of which the object is a point, and the second as a reference beam. When the hologram

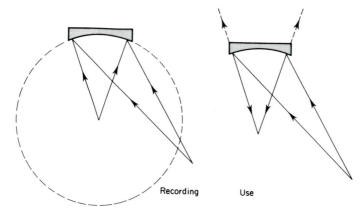

Figure 7.11 The construction and use of a holographic grating.

is illuminated with the reference beam it reconstructs the object beam but, since the grating is coated with a reflecting layer, the object beam is in effect reflected by the substrate so that the wavefront converges to a perfect focus at the centre of curvature of the blank. In holographic terminology it forms a real image of the original point source object. Of course this perfect image only occurs when the reference beam is in the same position as one of the recording sources and the wavelength is the same as that used to record the grating. In practice this condition is seldom achieved since the whole purpose of a grating is to study a range of wavelengths. In designing holographic gratings it is therefore necessary not only to achieve a perfect focus at one wavelength under one condition of illumination but to study the properties of the spectral image at different wavelengths and under the different conditions of illumination which correspond to various instrumental configurations. Before going on to discuss the way in which the full analysis of the problem is approached, let us first consider some simpler alternatives to the classical concave ruled grating.

One of the major features of a spherical ruled grating is its astigmatism. This arises because the effective radius of curvature (when compared to spherical mirror) is not the same in the plane of dispersion as it is in the plane parallel to the grooves. For a given object distance, the image distance is different in the two planes, and the vertical and horizontal focused images are separated. The focal conditions in the plane of dispersion are determined by Equation (7.3), but in the orthogonal plane the grating behaves like a simple curved mirror. It is therefore possible to correct the astigmatism by making a grating on a substrate that has different curvatures in the plane of dispersion and the plane orthogonal to it. The substrate is then part of a torus and it is possible to select the two radii of curvature so that at a chosen wavelength the vertical and horizontal image coincide to give a stigmatic focus. Gratings of this type have been ruled and do show the expected reduction in astigmatism (e.g. Strezhnev and Schmidt 1975). The problem is, however, that toric surfaces are not self-generating (i.e. if a surface and its replica are slid across each other and rotated they do not remain in contact at all points) and they are therefore far more difficult to make than spherical or plane ones.

If one has complete freedom over the spacing and shape of the grooves, then it is possible on a plane substrate to make a grating that will bring the diffracted light to a focus. From the point of view of holography, we can describe this as a hologram of a point formed with a spherical reference wave (Figure 7.12). Such a hologram consists of a series of concentric rings and closely resembles, at least on axis, the familiar Fresnel half-period zone plate. In the figure it is shown used in transmission, but it can, of course, also be used in reflection. The axial focal properties of a zone plate are well known: it behaves as a lens with a focal length of $r_0^2/m\lambda$ where r_0 is the radius of the central zone. The focal length is thus inversely proportional to the wavelength and the spectrum is formed as a series of images at different points along the axis of the system. This is not a practical arrangement as it stands because it is difficult to separate the focused spectral image at one point from the out of focus light at different wavelengths. However, if we make just an off-axis section of a zone plate we have something which rather more resembles a grating and this is no longer true.

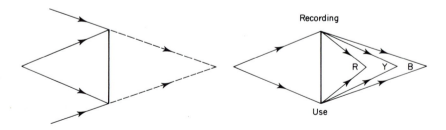

Figure 7.12 Chromatic dispersion of a holographic lens.

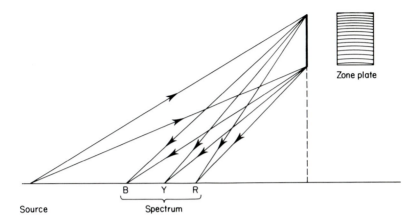

Figure 7.13 An off-axis section of a zone plate used as a focusing grating.

We see from Figure 7.13 that the focal curve is a straight line (along the axis of the zone plate) and this might be an advantage from the point of view of mechanical simplicity. Even so the aberrations that occur as one departs from the recording wavelength are so severe that such gratings are not in fact used in instruments. They do, however, illustrate the principle that light cannot only be dispersed by a grating but can also be brought to a focus by diffraction rather than reflection. The important difference between diffraction and reflection in this context is that reflection is achromatic, whereas diffraction is strongly wavelength-dependent (were it not so gratings would not be used as dispersing elements). As a rule it is therefore good practice to focus the spectrum as well as possible by using a substrate of the appropriate shape, and then use control over the shape and spacing of the grooves to reduce the residual aberrations. Owing to the difficulty in making aspheric substrates the usual practice is to use a spherical substrate and then to arrange the grooves so as to achieve the optimum correction of aberrations for a particular purpose. In principle it is possible to have complete control over the shape and position of the grooves by interfering wavefronts of the appropriate shape, but in practice the wavefronts are usually either spherical or plane. The process of designing a holographic grating then consists of setting up an expression for the differential of the light path function in the same way as for the conventional concave grating. In this case, however, the spacing and orientation of the grooves are not constant but depend upon the positions of the point sources used to record the grating. The coordinates of these points then find their way into the terms in the series expansion of the light path function and influence the focal properties and the various aberrations of the grating. By a suitable choice of the coordinates of the recording points, it is possible to optimize or even to eliminate particular aberrations which may be of concern for the application in which the grating is to be used.

Let us consider the basic equations of the holographic grating with reference to Figure 7.14. As before, the pole of the grating blank is at the origin of a set of cartesian coordinates (x, y, z). A, B and P are respectively points on the source, the image and the grating, and the light path function for this point is expressed as $F = \langle AP \rangle + \langle PB \rangle + nm\lambda$. However, in this case the expression for nm will depend upon the positions of the point sources C and D which are used to record the grating.

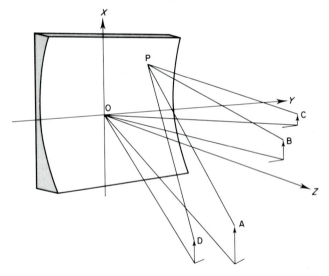

Figure 7.14 Parameters for the construction and use of a general holographic grating. $OA = r$, $OB = r_1$, $OC = r_{11}$, $OD = r_{111}$.

The positions of the grooves will be determined by the conditions of interference. That is, for a bright fringe the optical path difference between the two wavefronts shall be equal to an integral number of wavelengths, so that

$$N\lambda_0 = \langle CP \rangle - \langle DP \rangle$$

If we assume that there is a bright fringe at the origin then we may also write

$$n\lambda_0 = [\langle CP \rangle - \langle DP \rangle] - [\langle CO \rangle - \langle DO \rangle]$$

and this we can substitute into the expression for the light path function to obtain

$$F = \langle AP \rangle + \langle PB \rangle + \frac{m\lambda}{\lambda_0} [\langle CP \rangle - \langle DP \rangle - \langle CO \rangle + \langle DO \rangle]$$

(m is the order number and n is the number of grooves between the origin and P). This in turn can be expanded in just the same way as in the case of the classical grating. This expansion is rather lengthy so we shall write down here only the first few terms (Namioka *et al.* 1973):

$$
\begin{aligned}
F =\ & r\left(1 + \frac{x_a^2}{r^2}\right)^{1/2} + r_1\left(1 + \frac{x_b^2}{r_1^2}\right)^{1/2} \\
& - y\left[\left(1 + \frac{x_a^2}{r^2}\right)^{-1/2}\sin\alpha + \left(1 + \frac{x_b^2}{r_1^2}\right)^{-1/2}\sin\beta\right. \\
& \left. - \frac{m\lambda}{\lambda_0}(\sin\delta - \sin\gamma)\right] \\
& - x\left[\frac{x_a}{r}\left(1 + \frac{x_a^2}{r^2}\right)^{-1/2} + \frac{x_b}{r_1}\left(1 + \frac{x_b^2}{r_1^2}\right)^{-1/2}\right] \\
& - xy\left[\frac{x_a}{r^2}\sin\alpha + \frac{x_b}{r^2}\sin\beta\right] \\
& + \tfrac{1}{2}y^2\left[\frac{x_a^2}{2r^2}\left(\frac{2\sin^2\alpha}{r} - \frac{\cos^2\alpha}{r} - \frac{\cos\alpha}{R}\right)\right. \\
& \qquad + \frac{x_b^2}{2r_1^2}\left(\frac{2\sin^2\beta}{r_1^2} - \frac{\cos^2\beta}{r_1} - \frac{\cos\beta}{R}\right) \\
& \qquad + \frac{\cos^2\alpha}{r} - \frac{\cos\alpha}{R} + \frac{\cos^2\beta}{r_1} - \frac{\cos\beta}{R} \\
& \qquad \left. - \frac{m\lambda}{\lambda_0}\left(\frac{\cos^2\gamma}{r_{11}} - \frac{\cos\gamma}{R} - \frac{\cos^2\delta}{r_{111}} + \frac{\cos\gamma}{R}\right)\right] \\
& + \tfrac{1}{2}x^2[\qquad] + \ldots
\end{aligned}
$$

$$(7.23)$$

As one might very well expect by comparing Equation (7.23) and Equation (7.11), there is a great deal of similarity between the expansion of the light path functions for the classical grating and for the holographic grating. The various coefficients take on the same significance, but in each case in addition to the classical term there is a holographic term which depends upon the geometry used to making the grating. Thus, for example, in the classical case the term W_{01} led to the equation

$$\left(1 + \frac{x_a}{r}\right)^{-1/2} (\sin \alpha + \sin \beta) = \frac{m\lambda}{d}$$

which reduced to the familar grating equation

$$(\sin \alpha + \sin \beta) = m\lambda/d$$

where A is the meridional plane so that $x_a = x_b = 0$.
In the holographic case if we take

$$\frac{\partial F}{\partial x} = \frac{\partial F}{\partial y} = 0$$

and again set $x_a = x_b = 0$ we have

$$(\sin \alpha + \sin \beta) = \frac{m\lambda}{\lambda_0} (\sin \delta - \sin \gamma) \qquad (7.24)$$

which is again the grating equation except that the groove spacing is determined explicitly from the conditions of interference. As in the classical case the higher-order terms describe the conditions for horizontal and vertical focus and the various aberrations. In each case, however, the expression contains a classical part and a holographic part which is wavelength dependent and which is determined by the recording geometry. The process of designing a holographic grating then consists of choosing the recording parameters in such a way that the holographic part of the term representing a given aberration counteracts or cancels the classical term, so that the net aberration is substantially reduced. Which aberrations one chooses to reduce will be determined by the application for which the grating is required. However, once the recording parameters have been fixed to reduce a particular aberration at a particular wavelength, the holographic component of the other aberrations is also fixed and it may very well happen that by eliminating one aberration a different one is increased. For example, one might choose to reduce the astigmatism in a Seya–Namioka mounting only to find that this increases the amount of coma. So the design process is one of compromise and optimization and since the choice of optimum conditions varies from one application to another it is not possible to make a general comparison between the performance of ruled and holographic gratings. However, it is of interest to study Equation (7.23) in a little more detail and to consider the physical consequences of some of the solutions that are available.

First of all if we set $\gamma = \delta$ and $r_{11} = r_{111} = \infty$, then the grating is made with two collimated wavefronts which are equally inclined at the pole of the grating. All the holographic terms vanish and Equation (7.23) reduces to Equation (7.11), so the grating is equivalent to one ruled on a classical ruling engine. This is to be expected since, as shown in Figure 7.15, the interference of two collimated beams of light generates a series of parallel nodal planes in their common volume and the positions of the fringes are determined by the intersection of the grating blank with these planes. It is evident from the diagram that in this case the fringes will be equally spaced along the chord of the grating.

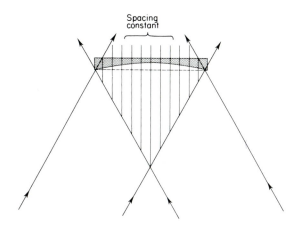

Figure 7.15 "Type I" holographic grating, equivalent to a ruled grating.

As in the classical case, the conditions of vertical focus are described by the coefficient of y^2 in the series expansion. If we consider only points in the meridional plane YZ, then we may write the focal condition as

$$F_{02} = 0 = \frac{\cos^2 \alpha}{r} - \frac{\cos \alpha}{R} + \frac{\cos^2 \beta}{r_1} - \frac{\cos \beta}{R}$$
$$- \frac{m\lambda}{\lambda_0}\left(\frac{\cos^2 \gamma}{r_{11}} - \frac{\cos \gamma}{R}\right) + \frac{m\lambda}{\lambda_0}\left(\frac{\cos^2 \delta}{r_{111}} - \frac{\cos \delta}{R}\right) \qquad (7.25)$$

One possible solution to this equation is that the terms in γ and δ are separately equal to zero so that

$$\frac{\cos \gamma}{r_{11}} = \frac{\cos \delta}{r_{111}} = \frac{1}{R}$$

which is once again the Rowland circle condition. This means in effect that if a grating is recorded using two point sources which both lie anywhere on the Rowland circle, the focal curve will be

$$\frac{\cos^2 \alpha}{r} - \frac{\cos \alpha}{R} + \frac{\cos^2 \beta}{r_1} - \frac{\cos \beta}{R} = 0$$

which is the same as for a classically ruled grating. However, we have also just seen that the same is true if the holographic part of F_{02} is made equal to zero by setting $\delta = \gamma$ and $r_{11} = r_{111} = \infty$, so we now have two entirely different recording geometries which give rise to the same focal curve. Obviously, these gratings are different: how is it then they they both have the same focal curve?

The answer is that the focal curve of Equation (7.25) refers only to a strip of grating of negligible height around the equator of the blank and that we have neglected the higher-order terms in the series expansion which describe the aberrations that become more important as the width of the grating increases. Within this approximation the spacing of the grooves is indeed constant along the chord in each case. On the other hand, the shape of the grooves will be different for the various recording geometries as indicated in Figure 7.16. So, although the positions of the spectral images will be the same for gratings of finite height, their shape will not. This is particularly important, because it means that it is possible to produce aberration-corrected holographic gratings that are compatible with classical ruled gratings and may therefore be substituted directly in instruments which were designed to accept conventional concave gratings.

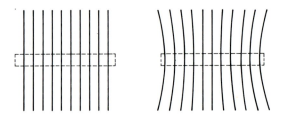

Figure 7.16 Different gratings having the same meridional focal curve but grooves of different shape.

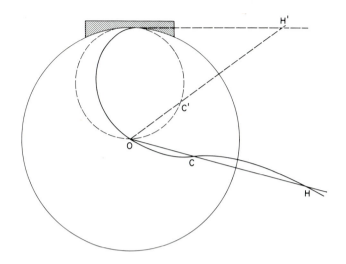

Figure 7.17 The focal curve of a holographic grating generated from point sources situated at harmonic conjugates with respect to the grating sphere.

On the other hand, if one is prepared to accept focal curves which do not conform to those available from the classical grating, then one has greater freedom to choose the recording parameters and therefore greater flexibility in achieving the desired correction of the aberrations. An interesting alternative series of focal curves was considered by Cornu in 1893 and in the light of holographic techniques has been studied in detail by Labeyrie, Flamand and Cordelle at the firm of Jobin Yvon in Paris. These gratings are made with two point sources, the positions of which bear a harmonic relation with respect to the grating sphere, and they have been called "Type III" gratings. (According to Jobin Yvon (1973) classifications, Type I refers to the grating made with plane wavefronts and Type II to a grating made with point sources on the Rowland circle, one of which is at the centre of curvature of the blank, as in Figure 7.11.) The relationship between the positions of the recording points and the grating sphere are shown in Figure 7.17, in which O is the centre of curvature of the blank and C and H are so arranged that

$$\frac{OC}{R} = \frac{R}{OH} = M$$

In principle C and H can be anywhere in the plane of dispersion, but if C is outside the grating sphere, H comes inside and the two exchange identities. If, however, C is on the Rowland circle H lies on the tangent of the sphere at the grating pole and if C is inside the Rowland circle then H is on the other side of the grating. Gratings made from two point sources situated at positions chosen from O, C and H have the interesting property that they are capable of forming stigmatic spectral images at *wavelengths other than the recording wavelength*. If the spectral source is positioned at any one of the points O, C or H then stigmatic images will be formed at each of the points O, C and H at wavelengths which depend upon the wavelength of manufacture, the positions of the recording sources and the harmonic ratio M.

The various stigmatic wavelengths corresponding to different positions of the recording sources and of the spectral source are summarized in Figure 7.18. In the matrices the top row corresponds to the position of the spectral source and the columns list the stigmatic wavelengths of images formed at the positions labelled in the first column. Thus, for example, in the case where the grating is recorded with point sources at O and C, if the spectral source is placed at C a stigmatic image of wavelength λ_0 will be formed at O (Figure 7.11 again!), there will be a stigmatic image at C of wavelength $2\lambda_0$ and one at H of wavelength $(M + 1)\lambda_0$. In this context the point sources used in recording may be either real or virtual. The significance of a virtual source is that the light is converging towards it rather than diverging from it. The significance of a negative value of wavelength in Figure 7.18(b) is that the spectral image is virtual.

Let us now consider in more detail the implications of Figure 7.18. The first point is that it is possible to choose the recording parameters so that the grating will form a stigmatic image at a chosen wavelength and that having done so there will then be other wavelengths for which a stigmatic image will be formed. If these wavelengths are chosen to be relatively close together then it is possible to reduce the astigmatism over a useful spectral range because, even though there will be a deterioration of image quality between the stigmatic wavelengths, the amount of astigmatism will usually be significantly less than it would have been with a classical concave grating. This is demonstrated in Figure 7.19, which compares the spectra formed with the two types of grating.

Aberrations other than astigmatism may also be introduced and these may tend to broaden the spectral image, but in practice this is unlikely to be a problem except when the highest resolution is called for. The price one pays for the increased flexibility is that

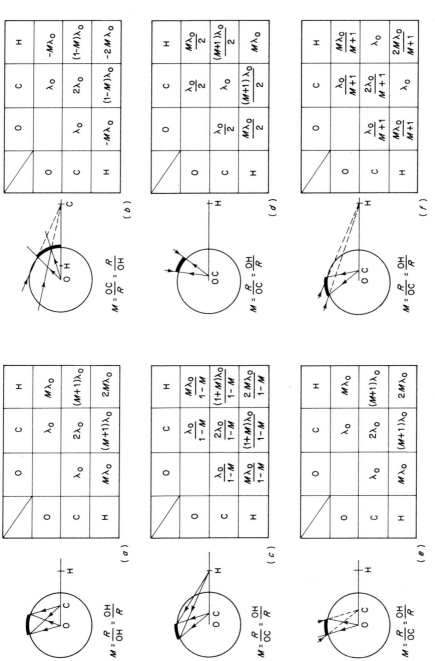

Figure 7.18 Stigmatic wavelengths for various forms of holographic grating.

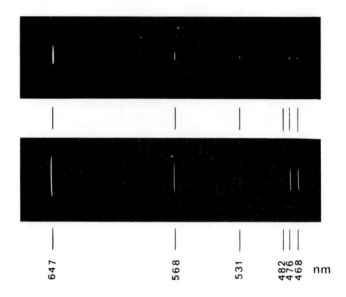

Figure 7.19 Comparison of a krypton laser spectrum from a conventional grating (lower) and a corrected grating (upper) showing reduced astigmatism.

the vertical focal curve is no longer a simple curve but a lemniscate (see Figure 7.17) which passes through the centre of curvature of the blank and the two harmonic points. (The horizontal focal curve is the straight line joining O, C and H.) Therefore, gratings of this type are not compatible with classical (or Type I) gratings and require special instruments to be designed to accommodate them. In designing such instruments one has the choice either of making a complex mechanism which will scan the spectrum and keep it in focus or adopting a simpler one and accepting a certain amount of defect of focus and the loss of resolution that goes with it. In many cases this is an acceptable solution and various instruments which are designed around Type III gratings are manufactured.

In each case described in Figure 7.18 a stigmatic focus may be obtained for a shorter wavelength by using a higher diffracted order, i.e. an image that is stigmatic for λ in the first order will also be stigmatic for λ/m in the mth order. However, this is not usually a satisfactory method of achieving stigmatism at short wavelengths, because the existence of many diffracted orders usually implies a low efficiency unless it is possible to blaze the grating. In fact, in the majority of the cases described in Figure 7.18, the recording conditions are such that a quasi-sinusoidal groove profile would be produced. Furthermore, as the value of m increases so, in most cases, the angle between the beams decreases and this places an upper limit on the pitch of the grating. Hence it can happen that in designing a grating for stigmatic focal conditions one excludes the possibility of using it under optimum conditions for achieving high efficiency (i.e. the single-diffracted-order configuration). We might also bear in mind that in the most straightforward of the configurations of Figure 7.18(a) and (c) the stigmatic wavelength is longer than the recording wavelength. Since the laser wavelengths which are available for generating interference gratings lie mostly between 250 nm and 500 nm, it follows that with these configurations it is not possible to achieve stigmatic focusing in the vacuum ultraviolet.

It is of course only in the vacuum ultraviolet that one really needs to use concave gratings at all. At longer wavelengths a plane grating with auxiliary mirrors could be used. There may from the point of view of stray light be some advantage in using concave gratings in the visible and near UV, and it may be that a monochromator with one concave grating can be manufactured more cheaply than one with a plane grating and two concave mirrors. In the vacuum ultraviolet a self-focusing grating becomes almost essential. Configurations (b), (d) and (f) offer the possibility of stigmatic focus at shorter wavelengths. In the cases of (b) and (f), a real stigmatic image is only achieved with a virtual point source. In other words, the light incident upon the grating is already converging to a focus; and since this implies the use of auxiliary optics we lose the advantage of using concave gratings anyway. This just leaves configuration (d) as a practical possibility. In this case, since $m > 1$, we can choose a stigmatic wavelenth down to $\lambda_0/2$. This does take us down into the vacuum ultraviolet provided that an ultraviolet laser (e.g. krypton at 351 nm or frequency-doubled argon at 257 nm) is used to make the grating. We shall see later that this is in fact a very important configuration, but its virtue lies not just in its holographic correction of aberrations but in the fact that it enables *blazed* concave gratings to be made.

The design of holographic gratings is a process in which the aberrations described by the various terms in the light path function may be controlled by adjusting the recording parameters. The less one has to constrain these parameters, the greater flexibility one has to reduce the aberrations. However, there are certain practical constraints. First of all, there is a limited range of wavelengths of manufacture for which suitable lasers are available. Secondly, since it is much more difficult to make aspheric blanks such as ellipsoids and toroids than it is to make spheres, one is often restricted to the use of spherical blanks. Thirdly, one may require that the focal equation be the same as that for a classical grating, in order to ensure compatibility with existing instruments. This imposes a further restriction on the recording parameters; one solution, though it is not unique, is to ensure that both recording sources lie on the Rowland circle. Fourthly, one may wish to specify the pitch at the pole of the grating. In this case one might for example find that it was possible to reduce either astigmatism or coma but not both, whereas both could be eliminated if the pitch were not specified. The geometries depicted in Figure 7.18 are, of course, only a selection of those available, as one has in principle complete freedom of the choice of the positions of the recording sources. This freedom permits the design of completely different types of grating, such as those with a flat focal field (for use in spectrographs) and those with a constant focal distance (for use with scanning monochromators). Although in these cases the instrument and the grating must be designed together, such instruments are available commercially and will no doubt have a significant influence on spectroscopy, especially at shorter wavelengths.

☐   Ruled gratings with corrected aberrations

The limited range of wavelengths available for recording "holographic" gratings can be restrict the range of possible designs. One way in which this limitation can be overcome is to revert to the mechanical ruling process with an engine that has been modified to rule curved grooves of varying spacing. Although interference techniques offer a convenient means of making aberration-corrected gratings, with a ruling engine one has in principle the ability to place the grooves wherever one wishes. Many ruling engines that are servo controlled could be adapted to give a variable spacing simply by inserting into the control loop a suitable phase shift for each groove and this could be controlled by a computer. Unfortunately, this technique cannot normally be applied to obtain curved grooves because if the deviation from straightness exceeds the grating spacing then the blank carriage will have to move backwards before the next groove can be ruled. This problem has been overcome by Harada and co-workers (1975) at the central research laboratories of Hitachi Ltd in Tokyo, who have built a ruling engine that not only provides for control of

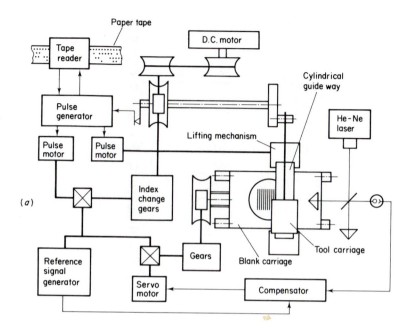

(a)

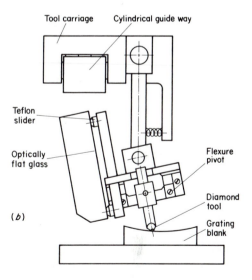

(b)

Figure 7.20 Details of an engine designed to rule aberration-corrected gratings. (Courtesy of T. Harada, Hitachi Ltd.)

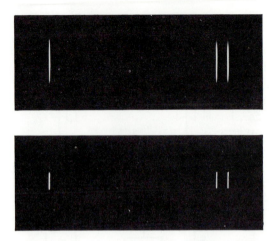

Figure 7.21 Examples of spectra obtained with the ruling engine of Figure 7.20: (*a*) conventional; (*b*) corrected. (Courtesy of T. Harada.)

groove spacing but will also rule curved grooves. Figure 7.20 shows a diagram of the control system of this engine and also details of the mechanism for ruling curved grooves. In this the ruling diamond is guided by a tilted optical flat and as the diamond moves up and down to follow the curvature of the blank so it is moved sideways to rule a curved groove. Figure 7.21 shows examples of spectra obtained from gratings ruled on this engine and demonstrates that it is indeed possible to design and rule stigmatic gratings without the limitations of laser wavelengths. One is perhaps tempted to refer to such gratings as "ruled" holographic gratings" thereby throwing into complete disarray the already confused terminology of this subject!

☐   Analysis of the focal properties using
        ray tracing techniques

In discussing the design of holographically corrected gratings we have concentrated mainly on the elimination or the reduction of the terms corresponding to aberrations in the series expansion of the light path function. This operation determines what one hopes are the optimum positions of the point sources used to record the grating. However, having done this it is advisable to check just how well the grating is likely to perform. A convenient way of doing this is to carry out, with the aid of a computer, an exact ray trace of light through the proposed system. By this means one can confirm that terms which may have been neglected in the original design are indeed negligible and can ascertain the image quality throughout the range of operation of the instrument. It is also a convenient means of studying the effect of straight or curved entrance slits of finite length and this will be particularly important if a point source had been assumed in the initial design.

In principle the process of ray tracing consists of considering a series of rays from the source to a grid of points on the grating surface. Where each ray meets the grating the local pitch and orientation of the grooves is calculated and the direction of the diffracted ray is determined by applying the grating equation as though at each point there was an elemental plane grating. The intersection of each diffracted ray with a chosen image plane (or surface) is then calculated and plotted with respect to coordinates in that plane so that a spot diagram of the image is built up. Strictly speaking one should consider a uniformly distributed bundle of rays emanating from each chosen point on the source and should calculate where these rays intersect the grating surface before pursuing the ray tracing programme. It is often simpler, however, to consider rays to an array of points which are regularly distributed on the grating surface; provided that the grating is not too steeply curved or the angle of incidence too great, this simplification will not unduly distort the weighting of the points in the spot diagram.

The basic ray tracing equation may, according to Welford (1975), be expressed very simply in the following vectorial form. In Figure 7.22 let n be a unit vector along the local normal to the grating surface at P, the point of incidence of the ray to be traced. Let r_r and r_0 be unit vectors along the rays from the two point sources used in the construction of the grating, let r_i and r_d be unit vectors along the incident and diffracted rays, and let λ_0 and λ be respectively the wavelengths of formation and use. For ray tracing it is required to determine r_d given r_i and the other data. The equation is

$$n \wedge (r_a - r_i) = \frac{m\lambda}{\lambda_0} n \wedge (r_0 - r_r) \qquad (7.26)$$

Where m is the order of diffraction, so that $m = 0$ refers to the ray reflected by the substrate.

If the vectors are resolved in a rectangular coordinate system orientated with its Z axis along the local normal, Equation (7.26) takes the form

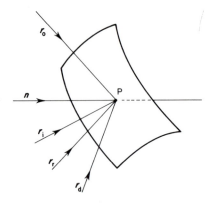

Figure 7.22 Parameters for the ray tracing of a holographic grating.

$$L_d - L_i = \frac{m\lambda}{\lambda_0}(L_0 - L_r)$$

$$M_d - M_i = \frac{m\lambda}{\lambda_0}(M_0 - M_r) \qquad (7.27)$$

where the components of a typical vector are (L, M, N). Equation (7.26) provides a simple linear solution for the components of r_d and it is interesting to note that the result is obtained without formally calculating either the pitch or orientation of the grooves.

This result, of course, merely takes the incident beam at the grating surface and determines the direction of the diffracted beam. Thus it corresponds to what is known as the "refraction process" in lens ray tracing. In the other half of the ray tracing process, known as "transfer", it is necessary to find where a ray emerging from one surface meets the next. This, however, is a simple matter of geometry and the procedures are well documented (Welford 1974).

□   The efficiency of concave gratings

So far we have only been concerned with the focal properties of concave gratings, but we must also bear in mind that making a grating on a concave substrate can also influence the efficiency. This is particularly true in the case of a classical ruled grating where a plane substrate is simply replaced by a concave one on a normal ruling engine. In this case the ruling diamond remains fixed and the facets of all the grooves are parallel so that, owing to the curvature of the blank, the angle of the facet with respect to the substrate varies across the aperture, as shown in Figure 7.23.

It follows that if the facet angle is correctly set for the principal ray then it will be incorrectly set for the rays striking other parts of the grating aperture. For simplicity let us consider the case of autocollimation, where the diffracted ray returns along the path of the incident ray. In this case the ideal blazed facet will be normal to the direction of the ray (at least in the scalar approximation which is adequate for our present considerations). Under these circumstances the groove facets should all lie on a series of concentric spheres centred about the common source/image position and separated by a distance $m\lambda/2$, where m is the order number. From Figure 7.23(b) it is clear that this condition is not fulfilled on a classical ruled grating, so that in practice, although the relative efficiency may be quite high at one position on the grating surface, it falls off as one departs from this point. The greater the numerical aperture of the system, the greater is the departure from ideal blazed conditions. However, we saw in Chapter 3 that other factors in the design of spectroscopic instruments suggest that their overall efficiency increases with numerical aperture. Thus there is a conflict between the desire to collect as much light into the system by having a large numerical aperture and the need to keep the numerical aperture low in order to maintain the grating efficiency.

Since the angle of incidence relative to the facet normal changes across the aperture, there will at each point effectively be a different blaze wavelength, which we can calculate in the following way. In Figure 7.23(c) the angle of incidence on the facet changes between points P and P$'$ by an angle of $\Delta\alpha$ equal to PSP$'$ which to a sufficient approximation is given by

$$\Delta\alpha = \tan^{-1}\left(\frac{\Delta W}{R}\right) \tag{7.28}$$

so that the angle of incidence at P is α and at P$'$ is $\alpha + \Delta\alpha$. The blaze wavelength at P$'$ is

$$\lambda_{\mathrm{B}'} = \frac{d}{m}\,[\sin(\alpha + \Delta\alpha) + \sin\beta] \tag{7.29}$$

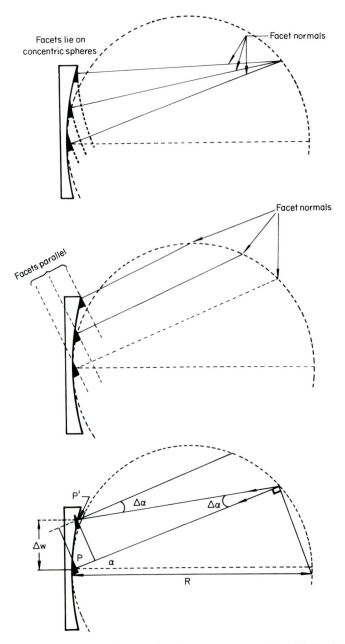

Figure 7.23 (*a*) The ideal facet angles for a concave grating; (*b*) the facet angles on a ruled grating; (*c*) parameters for the calculation of blazed wavelength at different points on a ruled grating.

$$\lambda_{B'} = \frac{d}{m} [\sin \alpha + \sin \beta + \cos \alpha \Delta \alpha]$$

$$= \lambda_B + \frac{d}{m} \cos \alpha \Delta \alpha$$

$$= \lambda_B \left[1 + \cot \alpha \frac{\Delta W}{R} \right] \tag{7.30}$$

To the approximation that $\Delta \alpha$ is small this result is valid for both Rowland circle and Wadsworth mountings. The practical result of this is that as one scans across the surface of the grating, the efficiency at a given wavelength varies; in fact it falls quite rapidly as one departs from the blaze peak. In order to counteract this it is often the practice to rule a concave grating in several sections and to adjust the orientation of the diamond between each section. Since it is not possible to do this without interrupting the continuity of the groove spacing there is an arbitrary phase difference between the various sections, and the resolving power of the grating will be that corresponding to the width of one section rather than the whole grating. It is common to rule such gratings in three sections, in which case they are referred to as "tripartite" gratings. The measurement of the variation of efficiency across the surface is best performed on an apparatus which illuminates a narrow area of the grating surface and in which the grating can be translated in its own surface, so that different areas of the grating can be illuminated without disturbing the geometry of illumination and detection. Such an apparatus has been built at the US Naval Research Laboratories and was described briefly in Chapter 5. Examples of the efficiency curves are shown in Figure 5.27. These show very clearly the rapid change of efficiency as one departs from the optimum blaze conditions and how this may be partially offset in a tripartite grating. They also show how the position of the blaze peak varies with wavelength and there is a good accord between the measured shift and that calculated on the basis of Equation (7.30) (Michels *et al.* 1974, Michels 1974).

This feature of the behaviour of ruled concave gratings puts into a different perspective the comparison between ruled and interference gratings. With a plane ruled grating one might expect a peak relative efficiency of 70% or 80%, which would offer a significant advantage over a symmetrical interference grating having at short wavelengths a typical relative efficiency of 20% to 30%. In concave gratings the advantage is less because although the peak efficiency from a symmetrical profile is lower, it does not change significantly across

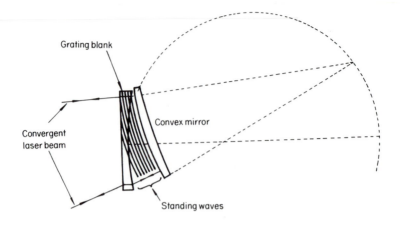

Figure 7.24 A method for making blazed concave interference gratings.

the aperture and it is therefore possible to operate with far larger numerical apertures without having to make the grating in different sections. Therefore the average efficiency for an interference grating with a symmetrical profile may not in fact be lower than that of a ruled blazed grating and one has the additional advantage that the "ruling" is coherent over the full aperture.

We saw in Chapter 4 that it was possible to generate blazed groove profiles in interference gratings in a variety of ways and in particular that by inclining the blank to a series of standing waves excellent gratings blazed for the ultraviolet were produced. This technique can be adapted for spherical gratings, as shown in Figure 7.24 (Hutley and Hunter 1981). This is an example of the type of holographic grating shown in Figure 7.18(d). The incident light converges to a point which in this case we have chosen to be on the Rowland circle. After passing through the grating blank it is reflected back along its own path by a spherical convex mirror, the centre of curvature of which lies at the virtual focus. A series of spherical standing waves is set up and these intersect the substrate to define both the position and shape of the grooves and their facet angles. In this case, however, the facet is always perpendicular to the direction of propagation so that the facet normals all point to the same point on the Rowland circle. The facets lie on a series of concentric spheres, which is just the condition required for a constant blaze wavelength over the whole grating surface. This view is in fact oversimplified because the facets are normal to the direction of propagation of the light within the bulk of the photo-resist and refraction at the resist/air interfaces causes the spherical standing waves to be tilted. As it happens, they appear to be centred around a different point very near to the Rowland circle and are closer together because of the effective reduction of wavelength inside a dense medium. This corresponds to a shorter blaze wavelength than that of the laser, as shown in Figure 7.25.

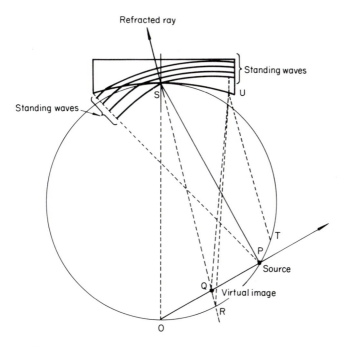

Figure 7.25 The effects of refraction on both blaze wavelength and the virtual image of the recording source.

This type of grating combines the best features of both ruled and interference gratings. It maintains the high peak efficiency of a saw-tooth groove profile over the whole of the aperture and it also offers the advantage of freedom from ghosts and grass. The range of blaze wavelengths that are available is well suited to concave gratings. If the recording is made directly in photoresist, then blaze wavelengths of 220 nm, 160 nm and 100 nm are available by using argon (459 nm), krypton or argon (351 nm) and frequency-doubled argon (257 nm), but these may also be reduced by ion beam etching to yield Littrow blaze wavelengths down to 20 nm if required. The uniform reduction of groove depth has the same effect as refraction (it reduces the separation of the notional spherical waves) so the facets are still centred near the Rowland circle and the blaze wavelength remains constant over the whole aperture. An efficiency map of such a grating is shown in Figure 7.26.

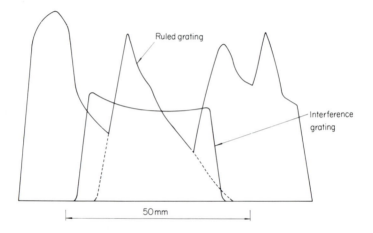

Figure 7.26 A comparison of the variation of efficiency across a blazed ruled grating and a blazed interference grating.

From the point of view of holographic reduction of aberrations, this type of grating is rather limited. It provides, as would be expected from Figure 7.18(d), a perfect focus for a point source of the wavelength of manufacture under the conditions of manufacture. It is recorded in effect with two coincident point sources, one real and one virtual. In order that the focal condition should be compatible with the Rowland circle mountings, the angle of inclination θ is given by $\lambda_0 = 2d \sin \theta$, where d is the required grating spacing at the pole and the distance from the pole to the virtual focus is $R \cos \theta$. The only parameters which are available to optimize the focal properties are then the laser wavelengths and the pitch and in many cases these are determined already by other factors. For a source placed at the position of the virtual source (i.e. autocollimation for λ_0) a stigmatic image is formed at the centre of curvature at a wavelength $\lambda_0/2$. There is therefore some improvement in image quality when used in the Eagle or Paschen–Runge mountings and if a UV laser is used to form the grating this improvement can extend well into the vacuum UV. However, in other mountings, such as the Seya–Namioka and the Wadsworth, there is no dramatic improvement over the focal properties of a classical ruled concave grating, but at the same time the image quality is not worse. The value of this type of grating in the UV is that it combines the high efficiency of the ruled grating with the low stray light and constancy of blaze of the interference grating. The blaze wavelength can be optimized for a chosen wavelength if necessary by ion beam etching to reduce the groove depth. However, the focal properties remain the same — that is, linked to the wavelengths of manufacture.

☐   The wavefront testing of concave gratings

The measurement of the quality of concave gratings by studying the diffracted wavefronts is far more complicated with concave gratings than with plane ones. In many cases, particularly with ruled gratings, the departure from the ideal (spherical) wavefront amounts to very many wavelengths of visible light and thus direct wavefront interferometry is not always very useful. Figure 7.27 shows an adaptation of a Michelson interferometer in which the focused wavefront from the grating is compared with that from a good concave spherical mirror. A cube beam splitter is used in order to equalize the aberrations of a spherical wave passing through a thick plate and a laser is required to provide adequate coherence as it is not usually possible to equalize the path lengths. The interferogram shows just how many fringes of error one has to contend with in a typical grating. Sometimes one wishes to study the quality of the rulings, in which case it is necessary to eliminate the aberration due to the blank. On other occasions it is the quality of the complete focused wavefront which is of importance.

One means of testing the quality of the rulings is to use shearing interferometry, in which two versions of the same wavefront interfere but one is shifted laterally with respect to the other (Birch 1966). This may conveniently be achieved by interfering two reflections from an inclined plane parallel block of glass as in the system shown in Figure 7.28. The light converges to its "focus", expands again and is collimated by a well corrected lens. If the grating is used in a Rowland circle configuration, such as the Eagle mounting as in the diagram, then most of the geometrical aberrations will be due to astigmatism and the collimated wavefront will be cylindrical

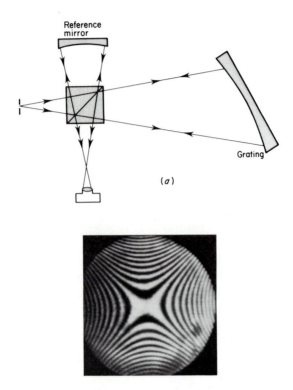

Figure 7.27 (a) An interferometer for the direct wavefront analysis of gratings; (b) interferogram of a conventional grating 0.15 n/a, 1200 grooves mm^{-1} at 633 nm.

rather than plane. By shearing the wavefronts along the axis of the cylinder, only the ruling errors are shown in the interferogram. It should be noted, however, that in this type of interferometry the sensitivity is less than in direct wavefront interferometry and depends upon the amount of shear introduced. It also requires the source to be a laser in order that the wavefronts should have adequate spatial coherence.

A visual assessment of the quality of the rulings can be made using a variation of the Foucault knife-edge test. This takes advantage of the fact that for a Type I ruled grating the image is stigmatic when formed at the centre of the Rowland circle in the Wadsworth mounting. Conversely, if a source is placed at this point the grating will act as a collimator. In the system shown in Figure 7.29 light is first collimated in this way and then reflected back along its own path by a plane mirror so that the grating brings it to a focus again.

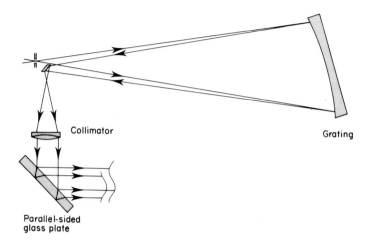

Figure 7.28 The use of a plane parallel plate to generate shear fringes. (After Birch 1968.)

A knife-edge is placed at the focus and ideally cuts out all the light. However, aberrations in the beam will cause light to miss the ideal focus and to go past the edge of the knife. If the grating is then viewed from behind the knife-edge, only those areas giving rise to errors in the wavefront will be visible. It is of course necessary that the wavelength should be less than the groove spacing, so for gratings of fine pitches one has to use ultraviolet light. In this case the light passing the knife-edge may be rendered visible using a vidicon image tube (Bruner 1972).

If we wish to study the quality of the focused image including geometrical aberrations, ruling errors and/or the effects of holographic correction, then we have, as in Figure 7.27, to compare the wavefront under test with a perfect spherical wavefront. An elegant means of doing this is the Linnick point diffraction interferometer, in which a partially transmitting screen is placed at the focus. In this screen is a small hole approximately the same size as the theoretical diffraction limit. Light passing through this is diffracted and forms a spherical reference wavefront which interferes with the rest of the wavefront that has passed through the screen. In order to achieve good contrast, the relative amplitudes in the two wavefronts must be adjusted by selecting an appropriate transmittance of the screen. This optimum will vary, of course, with the quality of the image, but in practice adequate contrast can be achieved over a wide range of values (Speer *et al.* 1979).

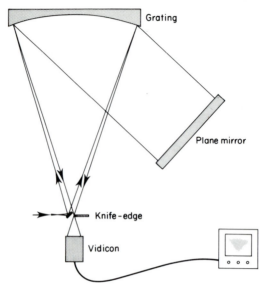

Figure 7.29 The knife-edge testing of a concave grating in the Wadsworth mount. (After Bruner 1972.)

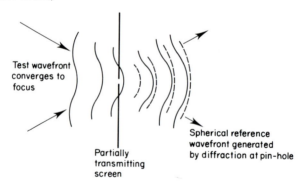

Figure 7.30 The Linnick point diffraction interferometer.

□   Summary

There are two distinct reasons why one might wish to use a concave grating. The first is that by combining the focusing optics with the dispersing element it is possible to design an instrument with fewer components than would be needed with a plane grating, and this would be cheaper to produce. Under these circumstances one is not usually concerned to achieve the ultimate in performance. For medium- and low-resolution instruments the compromise between image aberrations, shape of focal curve, slit curvature, efficiency and wavelength range is usually acceptable.

The second reason for using a concave grating is that at short wavelengths the extra losses caused by reflection at the collimating and focusing optics in a plane grating instrument are unacceptably high. This means that in order to pursue serious spectroscopy in the vacuum ultraviolet there is usually no choice but to use a focusing grating. Even with a conventional ruled concave grating the parameters have to be chosen with great care in order to achieve the best possible performance for a particular application. The radius of curvature, pitch, width and height have, as we have seen, all to be chosen with careful regard to the required resolution and dispersion and the maximum tolerable levels of slit curvature, astigmatism, spherical and other aberrations. The possibility of controlling the aberrations by varying the spacing and shape of the grooves has added a new dimension to the design of instruments incorporating concave gratings. It is possible to a large extent to tailor the properties of the grating to suit the application and in some cases to effect significant improvements in the performance of the instrument. The significance of these improvements will, of course, vary from application to application, so that it is impossible to give a general assessment of the value of the holographic reduction of aberrations. For example, if one wishes to photograph the spectrum of a point source, then a reduction in the astigmatism will give a far greater flux density in the image and the instrument will be correspondingly more efficient. On the other hand, if the same spectral image is detected photo-electrically behind a suitable exit slit, there may be little gained in reducing the astigmatism.

We cannot, therefore, summarize the results of comparisons between classical and holographically corrected gratings, but we can at least indicate how such a comparison can be made. The first stage is to design the holographic grating; this is done by selecting the recording parameters in such a way that the holographic terms in the expansion of the light path function counteract the classical terms for the particular aberration or combination of aberrations that one seeks to reduce. With the aid of a computer one next performs an exact ray trace to determine the shape of the image under the conditions that the grating will be used. From this one may calculate either the flux density or the flux passing through a slit and compare this with the equivalent calculations for a conventional grating.

There is no doubt that holographic techniques of making concave gratings can offer significant advantages both from the point of view of aberration corrections and of efficiency. The reduction of aberrations, particularly astigmatism, can be dramatic, but one is warned against assuming that there will necessarily be a correspondingly dramatic improvement is the instrumental performance. This can only really be judged after a careful analysis involving all relevant factors.

Bibliography

For further reading on the use of concave gratings, see:

R. A. Sawyer (1963). "Experimental Spectroscopy". Dover, New York.

For further reading on the design of holographic gratings, in addition to the references quoted in the text, see in the main reference list:

Noda *et al.* (1974)
Pouey (1974)

8

X-ray Gratings

□   Optics at grazing angles of incidence

Most of the features of gratings that we have so far described have
applied, at least in principle, to the gratings for the whole of the
electromagnetic spectrum. We saw in Chapter 6 that the behaviour
of a grating depends on ·the material of its surface, but the basic
form of gratings used for the infrared, the visible and the ultraviolet
is the same. However, as we go to shorter and shorter wavelengths
the reflectance of all known materials falls until at between 20 and
30 nm it is so low that it is practically impossible to record a
spectrum in a conventional instrument. At high frequencies, that
is with very energetic photons, the refractive index of all materials,
even the heavy metals, becomes very close to unity and so the
Fresnel reflection coefficient $R = [(n - 1)/(n + 1)]^2$ tends to zero.
In Figure 8.1 we show the familiar Fresnel reflection curves for gold
at a variety of wavelengths and we see that at short wavelengths
the only region in which the reflectance rises to a useful level is
for very high angles of incidence. For wavelengths below 20 or 30 nm
the whole of optics is dominated by the need to work at grazing
incidence and although in principle the same laws apply, in practice
the science of X-ray optics is very different from the optics of any
other part of the electromagnetic spectrum. It is a common, but
rather misleading practice, to use the term "normal incidence" when
referring to optics which is not at grazing incidence, so a grating used
in the Littrow mounting at an angle of incidence of $60°$ would be
referred to by X-ray spectroscopists as a "normal incidence grating"!

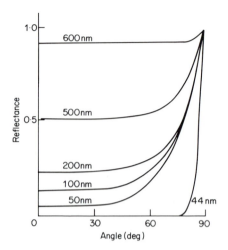

Figure 8.1 The reflectance of gold as a function of angle of incidence at various wavelengths.

Let us now consider some of the features which distinguish grazing incidence optics from conventional optics.

The spectral region for which grazing incidence optics is used extends from about 25 nm to less than 0.05 nm. This is the same dynamic range as from the far ultraviolet to near infrared and it includes short-wavelength ultraviolet, X-rays and γ-rays. However, for simplicity we shall adopt the common practice and refer to them collectively as X-rays. X-ray optics and in particular X-ray spectroscopy is dominated by two facts. First, for frequencies which are higher than the atomic eigenfrequencies of the material, the refractive index is close to, but slightly *less* than unity. The effects of refraction are so small that they are generally of no value in re-directing the radiation, but they are the reverse of those encountered in normal optics, so that, for example, a biconcave lens would cause X-rays to converge rather than to diverge. Furthermore, the total reflection of radiation incident at angles greater than the critical angle occurs outside rather than inside the medium and the phenomenon of total *external* reflection is the basis of most X-ray optics.

If we neglect the effects of absorption, the refractive index for X-rays of any material is given by the expression

$$n = 1 - \frac{Ne^2\lambda^2}{2\pi mc^2} = 1 - \delta \qquad (8.1)$$

where N is the electron density, λ is the wavelength, e and m are the charge and mass of the electron, and δ is of the order of 6×10^{-3} for gold at 5 nm. This approaches unity as the wavelength decreases, but in contrast to longer wavelengths it is less than unity. Snell's law states that, if α and γ are the angles of incidence and refraction then,

$$n = \frac{\sin \alpha}{\sin \gamma}$$

When $\sin \alpha > \sin \gamma$ there can be no refraction since this would correspond to a value of $\sin \gamma > 1$ and the radiation is therefore totally reflected. This is the basis for a great deal of X-ray optics, because it is the only practical means of attaining a useful level of reflectance. Since n is very close to unity it follows that the critical angle is very close to $90°$ and it is therefore usually more convenient to refer not to the angle of incidence α but to its complement, the grazing angle of incidence i. The critical (grazing) angle i_c for total external reflection is given by

$$\cos i_c = n$$

Since i is small we may write

$$1 - \frac{(i_c)^2}{2} = 1 - \delta$$

so

$$i_c = \lambda \left[\frac{e^2 N}{\pi m c^2} \right]^{1/2} = \lambda \cdot 3 \times 10^{-12} \cdot N^{1/2} \qquad (8.2)$$

when λ is in nm. λ_{min} is the shortest wavelength that may be reflected from a given material at a given angle of incidence, and may be expressed as

$$\lambda_{min} = \frac{i_c}{3N^{1/2}} \times 10^{-12}$$

From this we see that in order to reflect the shortest wavelengths we require the densest materials. Table 8.1 shows λ_{min} for a variety of materials with which X-ray optical components may be coated, either intentionally or by accident.

All of these values are based on the assumption that the surface is perfect and that absorption may be neglected. In practice this is not always so, and instead of the reflectance suddenly increasing to 100% as the critical angle is reached, the transition is more gradual. Figure 8.2 shows the measured reflectance of gold as a function of angle of incidence and from this we see that it is often necessary to

Table 8.1 Minimum wavelength reflected for a given grazing angle of incidence for various materials.

Material	Density	Number of electrons per cm^3	λ_{min} (nm)
Pentadecane (oil)	0.77	7×10^{22}	64.1 sin i
Glass	2.6	78×10^{22}	37.9 sin i
Aluminium	2.7	78×10^{22}	37.9 sin i
Aluminium oxide	3.9	115×10^{22}	31.2 sin i
Silver	10.5	276×10^{22}	20.1 sin i
Gold	19.3	466×10^{22}	15.4 sin i
Platinum	21.4	514×10^{22}	14.7 sin i
Iridium	22.4	542×10^{22}	14.3 sin i

to keep well below the critical angle in order to achieve high values of reflectance. It also shows that there is some reflection beyond the critical angle. However, in order to pursue our study of the design of X-ray gratings we shall assume that the grazing angle of incidence must be less than or equal to the critical angle, but shall bear in mind that in practice the total external reflection does not occur quite as the simple theory would suggest.

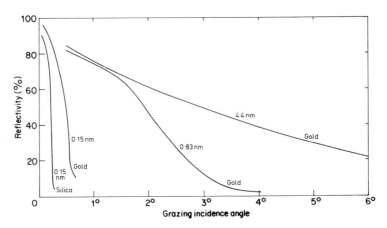

Figure 8.2 The measured reflectance of gold at X-ray wavelengths.

It has been suggested (Sprague *et al.* 1955) that for X-ray gratings the critical condition is not that of total external reflection but "total external diffraction", in which high efficiencies in the reflected orders of diffraction are obtained when the angle of incidence is such that the angle of diffraction for the first transmitted order would exceed 90°. There is evidence that diffracted orders are observed under circumstances where the conditions of total external reflection are not met. However, the matter is undoubtedly complicated by the fact that, just as at longer wavelengths, whenever a diffracted order skims the surface, anomalies are introduced and the efficiency in one order may be enhanced at the expense of another. Since all orders are within a few degrees of the surface, one is never far away from the condition under which anomalies are liable to occur. For laminar gratings, where the radiation is not simply reflected from the facets, it has been found appropriate to consider the reflectance corresponding to the mean of the angles of incidence and diffraction. At grazing incidence the diffracted orders are designated according to the scheme shown in Figure 8.3, in which negative orders lie between the zero order and the grating surface and positive orders lie outside the zero order.

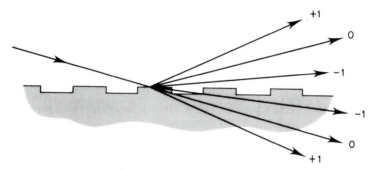

Figure 8.3 The labelling of diffracted orders at grazing incidence.

One factor which is most significant in many optical phenomena, particularly those involving diffraction, is the ratio of the wavelength and the physical size of the structure with which the radiation interacts. In the case of a grating the ratio λ/d plays an important role and often (though by no means always) it is possible to transfer results obtained in one spectral region to another simply by scaling up or down as appropriate. In practice diffraction gratings usually have a groove spacing between one and ten times the wavelength whether they are used in the infrared, the visible or the ultraviolet. If we were to apply the same conditions to X-ray gratings, we would be faced with the prospect of having to generate gratings with $100\,000$ grooves mm^{-1} or more. However, since it is necessary that the radiation be incident at small grazing angles, there is an apparent foreshortening of the features of the grating surface by a factor of the order of the sine of the grazing angle of incidence. For example at $1°$ a grating would be foreshortened by a factor of about 50 and 2000 grooves mm^{-1} would therefore be equivalent to 100 000 grooves mm^{-1} used at normal incidence; at a wavelength of 1 nm the equivalent value of λ/d is only 10. We saw from Chapter 2 that the greater the value of λ/d, the greater the dispersion of the grating and in Chapter 3 we saw that it is generally an advantage to have as much dispersion as possible. At grazing incidence this is no longer the case because in order to direct a detectable amount of energy into a diffracted order it is necessary that it emerge close to the critical angle. If the dispersion is too great then some of the spectrum may fall outside this angle and will be lost. Therefore, even for spectroscopy at wavelengths as short as 0.1 nm it is common practice to use gratings with between 300 and 2400 grooves mm^{-1}. These are pitches which are well within the capabilities of conventional grating technology to produce.

The foreshortening is illustrated in Figure 8.4(a) and applies to features which lie in the plane of the grating. Features which lie out of the plane of the grating are not foreshortened in the same way, but their effects on a reflected or a diffracted beam are reduced by a similar factor as shown in Figure 8.4(b).

Let us consider the reflection at grazing incidence from a surface with a small region that is a height h above the mean plane. At normal incidence the optical path difference between a ray reflected from that area and a ray reflected from the mean plane will be $2h$ and on the basis of the Rayleigh criterion may be neglected only if this is less than a quarter of a wavelength i.e.

$$2h < \frac{\lambda}{4}, \qquad h < \frac{\lambda}{8}$$

In the visible region this is about 50 nm. However, at grazing incidence the path difference is $2h \cos \alpha$ and therefore the tolerance on departures from perfection of the surface according to the Rayleigh criterion is now

$$h < \frac{\lambda}{8 \cos \alpha} \quad \text{or} \quad \frac{\lambda}{8 \sin i}$$

At the critical angle, $\lambda_{min} = K \sin i$, where K depends upon the material, and away from an absorption edge is given by $3.33 \times 10^{-13} \times N^{-1/2}$. We may therefore write:

$$h < \frac{K}{8} \text{ nm} \simeq 2 \text{ nm} \qquad \text{for gold}$$

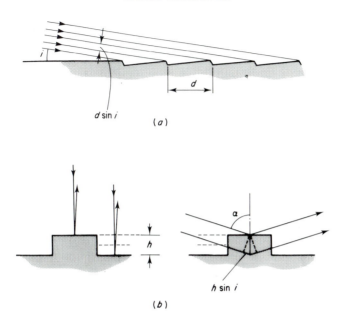

Figure 8.4 The foreshortening effect of features of the grating at grazing incidence.

This is to be compared with a tolerance in the visible region of about 50 nm. So, although the wavelength may be reduced by a factor of say 400, the tolerance on the optical surface has been reduced by only 25. This argument applies both to macroscopic imperfections that degrade the quality of the image and to microscopic imperfections that give rise to scattering and therefore to a loss of energy in the reflected beam. It is not possible at present to polish optical surfaces with a shape that is accurate to 2 nm, so the X-ray spectroscopist is unable to work at the theoretical resolving power. However, this is not a serious limitation because there are other factors which limit the resolution that is needed or can be achieved with X-ray instruments. In particular, the physical widths of slits that would be required to achieve theoretical resolution are very small, and the narrowest slit that can be made in practice is about 1 μm. A typical instrument might have an "f-number" of 100 and the slit width for theoretical resolution would then be 100λ. Therefore, for all wavelengths below 10 nm, the resolution would be slit-width-limited.

On the other hand, surface imperfections on a microscopic scale are very important because of the level of scattering that they generate. According to the theory of Bennett and Porteus (1961), the proportion of incident light that is scattered at near normal incidence due to surface roughness is given by

$$I = I_0 \left(4\pi \frac{h}{\lambda}\right)^2 \tag{8.3}$$

where h is the r.m.s. height of the roughness; h/λ is proportional to the phase difference between contributions from the roughness and from the mean plane, and at grazing incidence may be replaced by $(h/\lambda) \sin i$. So in order to reduce the scattering of X-rays to a level similar to that achieved in the visible, the degree of surface finish must be about 25 times better. Furthermore, since the scattering increases with the fourth power of the factor $(h/\lambda) \sin i$, it is extremely important to reduce the roughness as far as possible, otherwise a significant amount of energy will be lost to the signal and will contribute to the noise of the spectrum. The most important single factor which has influenced the progress of grazing incidence X-ray optics has been the development of techniques for producing "supersmooth" optical surfaces. In particular, at the National Physical Laboratory these techniques have enabled the value of the amplitude roughness (i.e. peak to valley) of vitreous silica surfaces to be reduced to as little as 0.5 nm with a corresponding improvement in the reflectance of mirrors and the efficiency of diffraction gratings (Lindsey and Penfold 1976).

☐   The design of X-ray gratings

So great is the influence of surface roughness that it plays a far more important part in determining the efficiency of an X-ray grating than does the shape of the groove profile. For this reason the development of X-ray gratings has been rather different from that of gratings used at "normal incidence" and the interest centred for a long time on groove profiles chosen because they could be made smooth rather than because the efficiency of the idealized profile was high. Three types of grating are of interest and these are represented in Figure 8.5.

The first, as shown in figures (a) and (b) is the grazing incidence equivalent of a bar-and-space amplitude grating, the second is a phase grating and the third is a blazed grating. We saw from Chapter 2 that

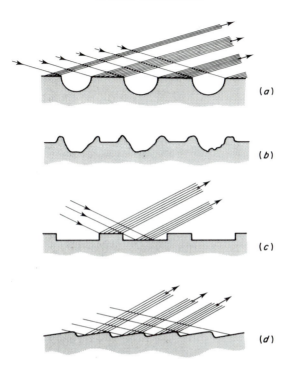

Figure 8.5 Various forms of X-ray grating.

the theoretical maximum first-order efficiency of an amplitude grating is about 10% and occurs when the bars and spaces are of equal width. It was with a grating of this type that the diffraction of X-rays by a ruled grating was first demonstrated by Compton and Doan in 1925. They used a grating with a pitch of only 50 grooves mm^{-1} which was lightly ruled in speculum metal and for the following 30 years or so X-ray grating spectroscopy developed on the basis of this type of grating. During this time the most celebrated source of such gratings was the ruling engine of M. Siegbahn in Stockholm (Siegbahn and Magnusson 1930, 1935). He usually ruled gratings not on metal but on glass. The main problem of ruling an amplitude grating is that the process does not remove material but merely displaces it, so that when a groove is ruled, material will be thrown up along the edges as shown in Figure 8.5(b). Not only does this cause significant scattering but it also casts a shadow on the smoother lands between the grooves. As a result of this the efficiencies of early gratings were often of the order of a fraction of one per cent. That is a factor of a hundred or so less than the theoretical value. Ruling in glass was often preferred to ruling in metal because the action of ruling can cause the open structure of glass to collapse locally. The displaced material then occupies a smaller volume and gives rise to less scattering and shadowing. With a metal the material is merely pushed to one side and forms ridges along the edge of the grooves.

Similar problems are encountered with the ruling of blazed gratings. With these not only do the burrs on the tips of the grooves give rise to severe problems, but the smoothness of the facets is determined by the ruling process and is often of very poor quality, as we see from the electron micrograph of Figure 5.17. In the case of a blazed grating it is the angle of incidence upon the facet rather than upon the grating that is of most importance, since in the geometrical optics approximation the radiation is simply reflected from the facet.

Each groove casts a shadow on the next one, as shown in Figure 8.6, so the relationship between the facet angle and the angle of incidence determines the proportion of the facet that is involved in the process of diffraction. If the angle of incidence on the facet is i_f and the blaze angle is ϕ and we assume that these are both small, then the proportion P of the facet which is illuminated is

$$P = \frac{B}{A + B} = 1 - \frac{\phi}{i_f} \tag{8.4}$$

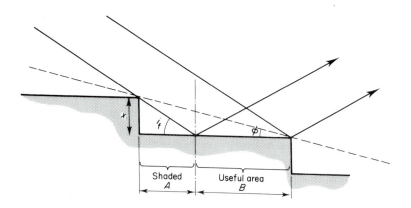

Figure 8.6 The effect of shadowing on a blazed grating.

From this it is obvious that the facet angle must be less than the critical angle and that it should be as small as possible in order to illuminate as large a proportion of the facet as possible. If only a small part of the facet is illuminated, then the performance of the grating depends critically upon the shape and texture of the tips of the grooves. This region is particularly susceptible to the effects of burrs and is very difficult to control on a ruled grating, although some advantage may be gained by using an odd-generation replica so that the tip of the groove is determined rather more by the shape of the ruling diamond. Figure 8.7 shows an electron micrograph of a ruled grating which was prepared for electron microscopy by shadowing at grazing incidence to the facets. The effects of the burrs can be seen clearly.

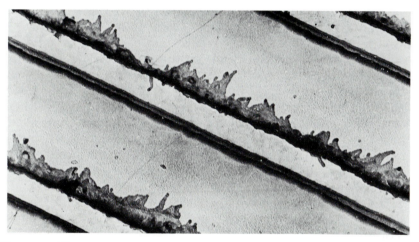

Figure 8.7 Electron micrograph of a ruled grating shadowed to enhance the burrs. (Courtesy of R. J. Speer.)

We saw in Chapter 2 that the blaze wavelength in a given configuration is related to the Littrow blaze wavelength by the expression

$$\lambda_B = \lambda_{b.\,Litt} \sin i_f$$

(Equation (2.18)), where i_f is the grazing angle of incidence on the facet. Since for a given surface it is, in effect, the value of i_f which determines the reflectance, this is the parameter which should normally be considered first when designing a blazed X-ray grating. Once this is chosen, then the optimum Littrow blaze wavelength may be determined and is independent of pitch. Thus if we wish to work at a wavelength of 4.4 nm, then from Figure 8.2 we might choose i_f to be about $2°$. The optimum value of the Littrow blaze wavelength would then be equal to $(4.4/\sin 2°)$ nm, or approximately 100 nm.

However, once the Littrow blaze wavelength is decided, the facet angle depends upon the pitch and is determined according to equation

$$2d \sin \phi = m\lambda_{b.\,Litt}$$

or in first order,

$$\sin \phi = \frac{\lambda_{b.\,Litt}}{2d}$$

Thus, the finer the pitch of the grating the greater is the facet angle and from Equation (8.4) the smaller the proportion of the facet that is illuminated. In our examples of a grating for 4.4 nm and $2°$ angle of incidence on the facet, gratings with 600, 1200, and 2400 grooves mm^{-1} will have facet angles of $1.7°$, $3.4°$, and $6.8°$ respectively. Only for the 600 grooves mm^{-1} grating would it be possible for the angle of incidence on the facet to be $2°$ and in this case the illuminated proportion of the facet is only 0.15. One can only design a finer grating to work at a greater value of i_f. In theory i_f must not exceed the critical angle so that this puts an upper limit to the pitch of the grating that can be used at a given wavelength. In the case of a material such as gold, in which the reflectance behaves as shown in Figure 8.2, it means that the higher the pitch the lower will be the efficiency. This is due to two factors: first the fact that the reflectance is reduced as i_f increases and secondly because the greater the pitch the smaller the proportion of groove that is used and the more sensitive the efficiency to defects of the tips of the grooves.

In order to express this more generally we may combine Equations (2.18) and (2.17a) in the following way:

$$\lambda_{b.\,Litt} = \frac{\lambda_b}{\sin i_f}$$

$$\sin \phi = \frac{\lambda_{b.\,Litt}}{2d}$$

$$\therefore \quad \sin \phi = \frac{\lambda_b}{2d \sin i_f} = \frac{\lambda_b}{2di_f} \tag{8.5}$$

for small angles.

The illuminated fraction of the facet is then

$$P = 1 - \frac{\phi}{i_f} = 1 - \frac{\lambda_b}{2di_f^2} \qquad (8.6)$$

and in the limiting case when only the tips are illuminated:

$$P = 0$$

and

$$d = \frac{\lambda_b}{2i_f^2} \qquad (8.6a)$$

In practice we would like to make P as large as possible, which implies that both d and i must be large, but in order to achieve high reflectance i should be small. Where there is a well defined critical angle this is the maximum value that i can take. Equation (8.6a) then sets a lower limit to the value of d and hence an upper limit to the groove density of the grating. If, on the other hand, in designing a grating for a given wavelength we specify at the outset the fraction of the facet that we wish to illuminate and choose the angle of incidence on this facet from a knowledge of the reflectance of the material, then the pitch of the grating is fixed by Equation (8.6).

For example, let us consider once again a gold grating working at 4.4 nm at an angle of incidence on the facet of $2°$ in which half the facet is to be illuminated. In this case

$$d = \frac{\lambda_b}{i_f^2} = \frac{4.4 \, \text{nm}}{(0.035)^2} = 3600 \, \text{nm} = 3.6 \, \mu\text{m}$$

so the grating pitch is just under $280 \, \text{grooves} \, \text{mm}^{-1}$. On the other hand, if for the same wavelength we were to double the angle of incidence on the facet, then the pitch could be four times finer, that is, just over $1100 \, \text{grooves} \, \text{mm}^{-1}$. Since, according to Figure 8.2, this would only entail a reduction of reflectance from 60% to 40%, such a choice would seem to be well justified.

The important feature of this analysis is that the design of a blazed X-ray grating is a compromise between, on the one hand, the desire to illuminate a large portion of the facet and to increase the groove densities, and on the other the need to reduce the grazing angle of incidence on the facet in order to maintain as high a value as possible of reflectance of the material. It also helps to explain the observed phenomenon that, at a given wavelength, gratings with high groove densities tend to be less efficient than coarser ones.

□   The manufacture of X-ray gratings

In practice ruled X-ray gratings, whether they are of the lightly ruled "amplitude" type or blazed, are significantly less efficient than one would expect from a knowledge of the measured reflectance of the material and the theoretical efficiency of the groove profile. The differences, which may be a factor of a hundred or more, can be ascribed almost entirely to the scattering and shadowing that arises from roughness in the grating surface. In order to produce efficient gratings it is essential that the surface be smooth and in order to obtain a smooth surface three conditions must be fulfilled: the substrate should be as smooth as possible; the formation of the grooves should introduce the minimum possible roughness; any reflecting coating should be as smooth as possible. During the period from about 1963 to 1975 significant improvement in the production of efficient X-ray gratings by Franks and his co-workers was made at the National Physical Laboratory by paying due attention to each of these three points (Franks *et al.* 1975).

In order to achieve the required degree of smoothness on the substrate it was necessary to develop special techniques of glass polishing. Substrates polished by hand on a conventional lap with regular relief (grooving) pattern on the pitch surface generally retained some "memory" of the polishing process and for best results it was necessary to use a polishing machine with an irregular relief pattern on the lap. The rate of removal of material is extremely important and this must be reduced to a minimum, particularly in the final stages of polishing. It is therefore easier to obtain a smooth surface on a hard material than a soft one. The homogeneity of the material is also very important in achieving a good polish. Vitreous fused silica is particularly suitable from this point of view and, since it also possesses good dimensional stability and a low coefficient of thermal expansion, it has been used extensively for the production of NPL X-ray gratings. The amplitude roughness of such "superpolished" surfaces may be as low as 0.5 nm and they may be used at wavelengths as short as 0.05 nm. At shorter wavelengths the reflectance is limited by roughness on an atomic scale. Having achieved a substrate with a smooth surface it is important to try to ensure that the surface of any reflecting layer is equally smooth.

The formation of evaporated thin films depends very strongly upon the material and the conditions of evaporation. Some metals, such as chromium are usually deposited uniformly while others such as gold and aluminium tend to form islands which conglomerate as

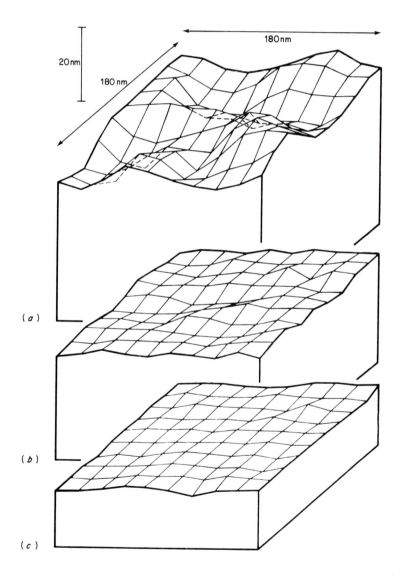

Figure 8.8 The surface topography of gold films measured by stereo electron microscope. Mean thickness: (a) 40 nm; (b) 29 nm; (c) 16 nm. (Butler 1973.)

the deposition proceeds. The depth of the film also plays an important part; Figure 8.8 shows a representation of the roughness of gold
films of different thicknesses deposited under identical conditions.
These results were obtained by stereo electron microscopy and
indicate that for best results the thickness should be 30 nm or less
(Butler 1973).

In order to avoid leaving debris on the grating surface it is
necessary that during formation of the grooves material should be
removed rather than merely displaced. It is necessary in some way
to etch the material and most etching techniques are more easily
applied to the generation of laminar groove profiles than to blazed
ones.

According to the scalar theory the maximum theoretical efficiency
of a laminar grating is $4/\pi^2$ or about 40%. This is for normal
incidence, the case in which the contributions from the tops and
the bottoms of the grooves are out of phase by π and cancel each
other in zero order. For this to happen the two components must
be of equal amplitude and therefore the bands must be of the same
width as the grooves. At grazing incidence the principle remains the
same, but other geometrical factors, particularly the effects of
shadowing, have to be taken into account. Three cases may be
considered and these are shown in Figure 8.9.

In the first case the effects of shadowing are ignored, the whole
of the grating area is illuminated and contributes to the intensity
of the diffracted radiation. Under these circumstances the maximum
theoretical efficiency is again $4/\pi^2$, but this of course is never achieved

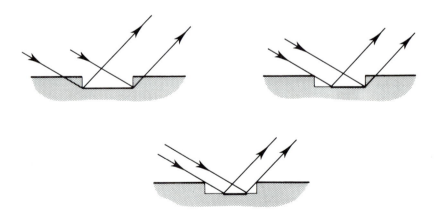

Figure 8.9 Shadowing effects on laminar gratings.

in practice. It does happen, however, particularly at very short wavelengths, that both the incident and diffracted radiation may penetrate the material supporting the upper land and the grating behaves as a phase grating even though the bottoms of the grooves should be completely shadowed. The second case is where the upper land does indeed cast a shadow to the incident radiation but where, since it emerges at a different angle, the diffracted radiation penetrates the grating. Finally we have the condition in which the upper lands both cast a shadow on the bottom of the grooves and absorb some of the diffracted radiation.

For a given angle of incidence on the facets, which in this case is equal to the angle of incidence on the grating, the effect of increasing the groove density is to decrease the proportion of the bottom of the grooves that is effective and hence to reduce the efficiency. So just as with blazed gratings, a finer pitch tends to imply a lower efficiency. However, in this case the limit is reached when none of the bottom of the grooves is effective and the grating becomes an amplitude grating. This still has a maximum theoretical efficiency of $1/\pi^2$ or 10%, in contrast to the completely shadowed blazed grating, where the efficiency is determined by the shape of the very tips of the grooves. In this respect the laminar grating has a theoretical advantage, although under these circumstances practical efficiencies tend to be influenced more by the ridges and burrs than by the theory.

In the case where masking is ignored, the relationship between the wavelength of the primary maximum of efficiency and the parameters of the grooves is given by the expression

$$\frac{\lambda}{d} = \frac{2 \cos i + (d/h) \sin i}{1 + (d/2h)^2} \tag{8.7}$$

where i is the grazing angle of incidence, d is the groove spacing and h is the groove depth which, if i is small and $h \ll d$, enables us to express the optimum groove depth as

$$h = \left(\frac{\lambda d}{8}\right)^{1/2}$$

The (scalar) theoretical efficiency for a fixed angle of incidence is shown in Figure 8.10 as a function of the normalized wavelength λ/d for a variety of groove depths and indicates that by a suitable choice of groove depth a laminar grating may be designed to operate over a chosen range of wavelengths. Where masking occurs it is usually preferable to have the grooves rather wider than the lands in order to equalize the contribution from both surfaces, thus

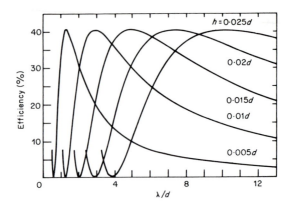

Figure 8.10 Theoretical positive first-order diffraction efficiencies as a function of wavelength for laminar gratings of various groove depths, h. Grazing angle 3°, unmasked case.

achieving more complete destructive interference in the zero order and the optimum efficiency in the first odd diffracted orders. The optimum ratio of the widths will of course depend upon the angle of incidence, the pitch and the groove depth.

In the manufacture of laminar gratings it is necessary to generate a mask which will protect the lands while the grooves are being etched and which may then be removed to reveal the smooth lands. The mask will then consist of a series of parallel stripes, but the form which it takes will depend upon the means used to etch the substrate. Two types of etching have been used for making laminar gratings, simple chemical etching and ion-beam bombardment. Chemical etching is simpler and cheaper and may well be faster, but tends to leave a rather rougher surface at the bottom of the groove than does ion etching. Figure 8.11 summarizes a variety of techniques which have been used to generate X-ray laminar gratings.

For coarse gratings with fewer than about $200\,\text{grooves}\,\text{mm}^{-1}$, it is possible to use conventional photolithographic techniques in which the groove pattern is contact-printed from a metrological grating onto a layer of photoresist, as in Figure 8.11(a). For finer gratings contact printing becomes very difficult because of the difficulty of maintaining intimate contact between the master and the blank.[†]

[†] Finer gratings can be made by contact printing using conformable masks or by using X-ray lithography. However, these techniques do not lend themselves to the production of spectroscopic gratings.

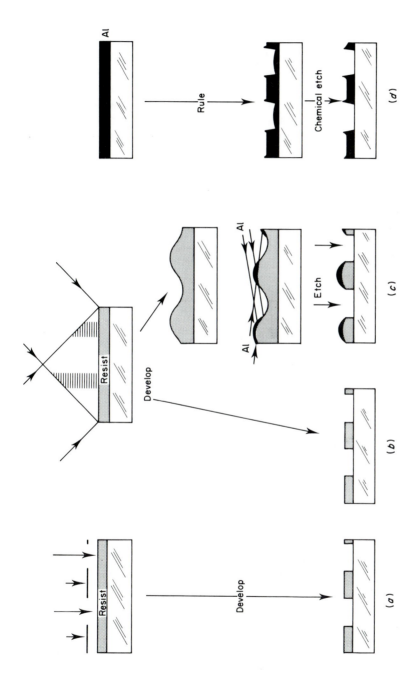

Figure 8.11 Methods of generating masks for the preparation of laminar gratings.

However, a similar result may be obtained using interference techniques, as shown in Figures 8.11(b) and (c). In Figure 8.11(b) the photoresist is exposed to the fringe pattern and developed right through to the substrate; in (c) a thicker layer of resist is used and this is developed to the shape of a surface undulation. The tips of the grooves are then coated with aluminium, which acts as a mask, and the photoresist is etched by ion bombardment at a much faster rate than the aluminium on the substrate. This is a somewhat cumbersome technique, but it does not require the same degree of control over the exposure and development of the photoresist that is required by method (b) and it does provide, by suitable choice of the angle at which the aluminium is deposited, a better degree of control over the bar-to-space ratio.

From an historical point of view, one of the most important techniques was that developed by Franks and Lindsey and which is shown in Figure 8.11(d). They started by ruling a grating with a symmetrical profile in a layer of aluminium on a "supersmooth" substrate. In order to avoid introducing stresses in the surface of the substrate, the rulings did not penetrate the aluminium and the grooves were then etched chemically in order to expose the substrate. The remaining aluminium strips then acted as a protective mask while the grooves were etched by ion bombardment into the substrate. Using this technique they were able for the first time to produce 300 or 600 grooves mm^{-1} gratings with really smooth profiles and to demonstrate the significant improvement in efficiency that can be achieved when surface roughness is reduced to a minimum. Although this process for making the protective mask has now been superseded by interference techniques, the basic idea remains the same. This is that the lands are protected during the whole process and so retain the smoothness of the original substrate. The grooves are then etched to a controlled depth, in this case by ion bombardment, in such a way that the troughs are also as smooth as possible and are therefore able to contribute to the process of diffraction.

Interference techniques for generating the mask have the advantages that they are faster, that it is possible conveniently to work with finer pitches, and that gratings may be applied to substrates which are too steeply curved to rule. They also make possible a variety of ways in which the mask may be used to generate a laminar profile and these are summarized in Figure 8.12. The first process shown is that developed by Franks and Lindsey in which the substrate is etched, the mask is removed chemically and the whole grating is coated with a metal layer (usually gold) to enhance the reflectance.

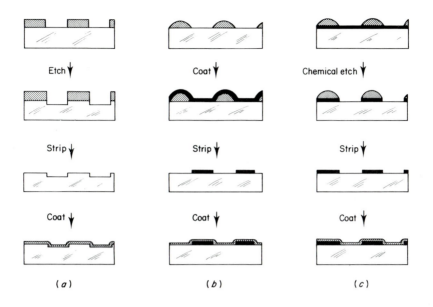

Figure 8.12 Methods of creating a laminar groove profile from a suitable mask.

In this case it does not matter which type of mask is used provided that it is able to protect the substrate during etching, and the same efficiencies have been recorded with both ruled and interference gratings.

With a mask of photoresist it is possible to generate a laminar profile by evaporating over both the resist and the bare substrate a layer of chromium (Rudolph *et al.* 1973). This is sufficiently porous that the underlying photoresist may be removed chemically. That chromium which is coated over the resist is also removed but that which is deposited directly on the substrate remains to provide the lands of the laminar profile. Theoretically this method has the advantage that the groove depth is determined by the thickness of the evaporated layer and that this may be controlled more accurately than the depth of a chemically etched groove. Unfortunately, the process of stripping tends to cause burrs at the edges of the lands and these cast shadows which severely reduce the efficiency of the grating. Furthermore, if the resist is not completely removed from the bottoms of the grooves, it will scatter radiation and also reduce the efficiency.

In the third method the mask is prepared on an evaporated layer of a material such as copper, the thickness of which is made equal to the depth of the required groove (Johnson 1975). This may then be etched chemically to reveal the smooth surface of the substrate at the bottom of the grooves. When the mask is removed, the smoothness of the lands is determined by the quality of the evaporated film; there are no burrs and efficiencies have been measured which are much the same as those of gratings produced by ion etching.

Although blazed gratings made by ruling with a diamond are usually less efficient and scatter more than laminar gratings, the sawtooth profile is still theoretically more efficient than the laminar one. The techniques of making blazed interference gratings may be adapted to produce blazed gratings that are suitable for use with X-rays. As we saw in Equation (2.18), when the grating is to be used at grazing incidence the effective blaze wavelength is significantly reduced compared with *Littrow* blaze wavelengths, and therefore gratings blazed for between 100 nm and 20 nm at "normal incidence" are suitable for X-ray use. We saw in Chapter 3 that using the Sheridon technique it is possible to obtain in photoresist, gratings blazed for wavelengths between 220 nm and 100 nm depending upon the wavelength of the laser. In order to make such a grating suitable for X-ray use, the slope of the facets must be reduced by a factor between 3 and 10 and this can be achieved using ion-beam bombardment etching. The method takes advantage of the fact that photoresist and silica will be etched at different rates, and was described in Chapter 4 (Figure 4.23(d)). By this means blazed X-ray gratings have been made with efficiencies that equal and at best are somewhat better than laminar gratings, particularly at higher spatial frequencies (Stuart *et al.* 1976). As we have seen, however, at the higher spatial frequencies the performance is determined by a smaller proportion of the facet, so that it is very susceptible to any "rounding off" of the tips.

☐   The performance of X-ray gratings

A comprehensive survey of the efficiencies of gratings for grazing incidence use was undertaken at the grating measurement laboratory in the physics department of Imperial College, and is updated as new gratings become available. The apparatus for these measurements was shown in Figure 5.28 (p. 173) and is capable of accepting either plane or concave gratings of a wide range of radius of curvature.

The absolute efficiency is determined by first measuring, with the grating removed, the intensity of monochromatic incident radiation. The detector is then scanned through the various diffracted orders to measure not only the efficiency in the orders but also the level of scattered light in between. This process may be repeated for various angles of incidence and various wavelengths so that a complete picture of the grating's performance may be built up. A series of results representing different types of grating is shown in Figure 8.13.

The bottom half of Figure 5.28 shows a typical scan for a ruled laminar grating at a single angle of incidence. The intensities are plotted on a logarithmic scale so that the incident beam, the diffracted beams and the stray light may all be shown. Figure 8.13 on the other hand, shows the efficiencies as a function of angle of incidence for a replica of a ruled grating, an ion etched laminar grating and a blazed interference grating. It is interesting to note that although the first-order efficiencies of the laminar grating and of the blazed interference grating are approximately the same, the shapes of the curves are rather different. In particular the existence of a minimum in the zero order curve for the laminar grating indicates that there is phase cancellation of the contributions from the top and the bottom of the grooves. The efficiencies in the positive and negative first orders are approximately the same for the laminar grating but are markedly different for the blazed grating and this confirms that the behaviour of the different types of grating are in general accord with the theory. The efficiencies are still rather lower than would be expected and this discrepancy must be attributed to surface imperfections and lack of control over the fine detail of the groove profile.

Figure 8.14 shows a summary of the efficiencies of X-ray gratings measured at Imperial College. It illustrates the dramatic improvement in efficiency that has taken place during the past 15 years, mainly due to the care that has been taken to reduce burrs and surface roughness. The highest efficiencies that have so far been measured have been those of blazed interference gratings which have been etched into a silica substrate.[†] However, as we have seen, there are some limitations to the extent to which blazed gratings may be applied and, particularly at high frequencies, the performance

[†] At 300 grooves mm^{-1} there are no data on blazed interference gratings. The "record" (1980) is held by an excellent blazed ruled master grating.

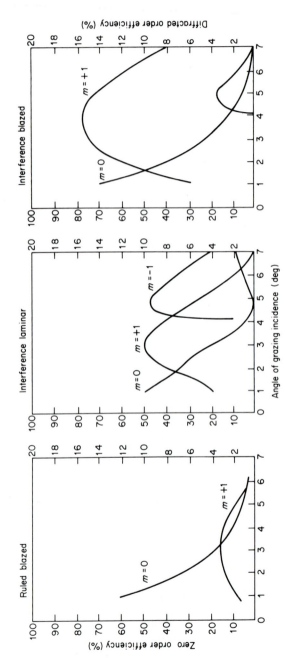

Figure 8.13 Efficiency measurements for ruled, laminar and blazed interference gratings.

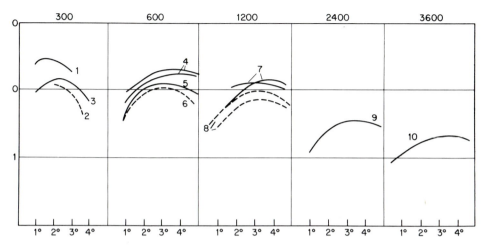

Figure 8.14 Summary of the best efficiency results at 4.5 nm arranged according to grating frequency.

1. Blazed ruled
2. Laminar holographic
3. Laminar holographic
4. Blazed holographic
5. Laminar ruled
6. Laminar holographic
7. Blazed holographic
8. Laminar holographic
9. Blazed ruled
10. Blazed ruled

depends critically on the shape of the tip of the groove. This is difficult to control, and efficiencies are therefore not always higher than those of laminar gratings. Furthermore, since the distribution of energy between the various orders is different, the choice between a blazed and laminar profile may well depend upon the application for which the grating is required. A second feature illustrated by Figure 8.14 is that the efficiencies are lower at higher spatial frequencies. This we have explained to some extent in terms of shadowing and of illuminating a smaller proportion of the groove. As it is much more difficult to measure the profile of finer grooves, it is also more difficult to control their shape and this too may contribute to the lower efficiencies. Whether the reduction of efficiency is offset by the increase of dispersion is, of course, a matter of instrumental design and once again depends upon the way the grating is to be used.

☐   Concave gratings at grazing incidence

The reflectance of mirrors at X-ray wavelengths is often low and there is considerable incentive to use concave gratings in spectroscopic instruments at grazing incidence. Plane gratings are sometimes used with mirrors particularly in conjunction with intense sources such as electron synchrotrons, but in general it is necessary to use a concave grating. For spherical gratings the Rowland circle condition applies as it does on the "normal incidence" domain, but owing to the obliquity of the incident and diffracted wavefronts the aberrations are very pronounced. A typical Rowland circle configuration is shown in Figure 8.25, from which it is evident that the astigmatism is almost complete and for a spherical grating the focusing in the vertical plane can to all intents be ignored. For example, the length of the astigmatic zero order image of a point source formed by a grating of 5 m radius and 20 mm height used at a grazing angle of incidence of $2°$ is 39.9 mm, whereas if the grating had been a vertical cylindrical surface with no attempt at focusing in the vertical plane, then the image would have been 40 mm. Astigmatism therefore causes a significant loss of intensity in the spectral image which is very serious for spectrographs. By itself though, it does not impair resolution of the instrument, so it is less of a problem for photoelectric spectrometers.

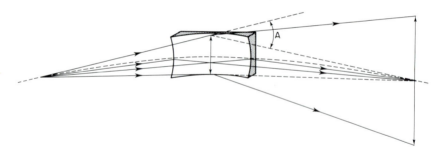

Figure 8.15 Astigmatism in a concave grating at grazing incidence.

As we saw in Chapter 7 there are two possible approaches to the problem of reducing the aberrations of concave spherical gratings; by forming the grating on a suitable aspheric substrate or "holographically" by varying the shape and spacing of the grooves. At longer wavelengths the holographic approach is found to be quite effective, particularly in the reduction of astigmatism, and the difficulty and expense of generating aspheric surfaces is often not justified. At grazing incidence, on the other hand, the scope of the holographic method is limited by the fact that it is not possible to deflect X-rays efficiently through large angles. For example, in

Figure 8.15, the angle through which the ray labelled A must be deviated in order to bring it to a stigmatic focus may well be much greater than double the critical angle. So although it would undoubtedly be possible to design a pattern of grooves such that the required diffracted order would be brought to a perfect focus, the efficiency away from the meridional plane may well be so low that there is no net advantage. It is therefore not practicable to eliminate astigmatism in this way and it is much more profitable to form the grating on a suitable aspheric substrate. The ideal shape of substrate to provide a perfect real image of a point source is an ellipsoid of revolution in which the source and image are placed at the two foci of the ellipse. For gratings the same function is best performed by a toroidal surface, because it may be used over a wider range of wavelengths, whereas with an ellipsoid a different sector would be required for each. Often, however, over the limited aperture that is required for an X-ray grating, the two surfaces are, to within normal glassworking tolerances, the same. The advantage of the toroid is that it has two distinct but constant radii of curvature and is therefore easier to make. Even so the techniques of generating aspheric surfaces are far from simple and the difficulties are compounded by the need to produce the smoothest possible surface. Furthermore, at grazing incidence the difference between the two radii of curvature is very great so that the blank more resembles a cylinder than a sphere. However, in practice neither the ellipsoid nor the toroid offers a complete solution to astigmatism because with a source of finite size the off-axis aberrations are very significant.

Interference techniques are well suited to the manufacture of gratings on unusual substrates. The first toroidal grating of this nature (shown in Figure 8.16) was made jointly at Imperial College London and the University of Göttingen (Speer *et al.* 1974). The radius of curvature in the meridional plane was 2 m and in the sagittal plane 5.65 mm. This grating was made by employing the techniques described in Figure 8.10(*b*) and 8.11(*c*), using straight equidistant fringes, and the area of the diffracted image was estimated to be about 35 times smaller than that which would have been obtained from a spherical grating. Furthermore, detailed ray tracing studies indicate that by a suitable choice of the parameters this improvement may be considerably increased. In a spectrograph in which a point source is to be imaged on to a photographic plate (as, for example, when studying the emission from a deuterium target in laser nuclear fusion experiments) this would represent a gain of 35 in the efficiency of the instrument and this is obviously

Figure 8.16 Toroidal grating for use in the soft X-ray region; primary radius 2000 mm, secondary radius 5.65 mm. (Courtesy of R. J. Speer.)

a far more important gain than could ever be achieved by improving the efficiency of the grating. The gain is, however, rather less when the grating is used in an instrument with slits. It is worth noting of course that this grating would have been impossible to produce using conventional ruling techniques.

The use of a toroidal surface introduces more spherical aberration than is associated with a sphere, but spherical aberration may, in principle, be corrected by a suitable variation of groove spacing. Since the angular deviation that one seeks to introduce is generally less than that already introduced by the grating, the "holographic" correction of spherical aberration is more practicable than that of astigmatism. Even so, the variation of pitch that is called for is rather extreme. Turner and Speer (1974), for example, have proposed a simple law determining the variation of pitch which is required to bring all rays to a perfect focus. According to this it would be possible to design gratings with far greater numerical aperture than would otherwise be possible. For example, a grating of 2.5 m radius of curvature and 200 mm long could be designed, but the grating frequency may vary from about 180 to 2200 grooves mm^{-1} across the aperture, which would pose serious problems in manufacture. It is unlikely that such a pattern of grooves could be formed by interference fringes, but it might be possible to rule it on a computer-controlled ruling engine such as that developed by Harada and discussed in Chapter 7.

Bibliography

A Franks (1977). *Sci. Prog.* **64**, 371.
R. J. Speer, (1976). The X-ray diffraction grating, *Space Science Instrumentation* **2**, 463.

Nonspectroscopic Uses of Diffraction Gratings

□ Introduction

So far we have been concerned almost exclusively with diffraction gratings that are intended for use as dispersing elements in some form of spectroscopic instrument. It was for this purpose that gratings were mainly developed and today it is the use to which most gratings are put. However, there are other uses for gratings and in this chapter we shall consider some of them briefly. It is not our purpose to give a detailed description of these applications but merely to point out that they exist and that spectroscopy is only one of the uses for gratings. Often the gratings themselves are in a rather different form from those we have so far described, but in many cases they have either been derived from spectroscopic gratings or have benefited from the technical advances inspired by the needs of spectroscopy.

One of the most important, and certainly the best established, alternative uses for gratings is as scales for the measurement of linear or radial displacement. Techniques for using gratings in this way were developed in the 1950s by Ferranti Ltd at Edinburgh using gratings which had been produced at the National Physical Laboratory following the work on the Merton–NPL process for making infrared gratings. The Merton process itself is not now used for making metrological gratings but it is of historical significance because it provided the first source of gratings suitable for metrology. Grating measurement systems are now widely used in the machine tool industry, both on measuring machines and on numerically

293

controlled machines. They provide a high degree of accuracy because the scale is independent of the machine drive so the effects of back-lash, wear and distortion are reduced. (These advantages are similar to those of controlling a ruling engine with an interferometer rather than relying on the screw.) A further advantage is that the infor-mation is provided in a digital form which is particularly suitable for electronic processing.

There are several ways in which gratings may be used as measuring scales and these have been described in some detail by Guild (1956, 1960) and by Rosamova and Gerasimov (1963). We shall consider some of the more commonly used techniques. The principle of operation is shown in its simplest form in Figure 9.1, in which two gratings are superposed; each is of the form of a series of alternate parallel opaque and transparent strips of equal width and they are aligned with the lines parallel. When the gratings are aligned so that the opaque strips are in register, light will be able to pass through transparent regions as in (a). When one grating moves relative to the other by a distance $d/2$, equal to half the period, then the opaque regions of one grating obscure the transparent regions of the other and no light is transmitted. As one grating is moved past the other the transmission varies periodically and by counting the number of oscillations it is possible to measure the distance moved. The sensitivity of the method is not limited to the period of the grating, because it is possible to interpolate a small fraction of a cycle of the signal.

In use the system consists of a long grating which acts as the

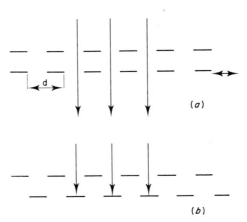

(a)

(b)

Figure 9.1 The superposition of metrological gratings: (a) in phase, (b) out of phase.

measuring scale and a reading head which consists of a smaller "index" grating together with a light source, a detector and the electronics necessary for generating and counting pulses as the head moves with respect to the scale. In addition simply to counting the pulses it is necessary to determine the direction of motion so that errors due to vibration or inadvertent reversal of the direction of the motion may be avoided. This may be achieved by recording not just one signal but two, each with a different phase, and these may be generated either by having a composite index grating where different regions have different relative phases or by rotating the index grating slightly with respect to the measuring grating. When two identical linear gratings are superimposed at a slight angle, one observes a pattern of moiré fringes consisting of parallel lines perpendicular to the bisector of the directions of the grating lines. Such a fringe pattern is shown in Figure 9.2. The spacing is $d/\sin \theta$ and as one grating moves by a distance x perpendicular to its lines so the moiré fringes move by a distance $x/\sin \theta$.[†]

If photocells are positioned across the width of the index grating as shown, then as the head moves each will generate a signal which is approximately sinusoidal and by a suitable choice of the positions of the detectors the signals may be arranged to be out of phase by $\pi/2$. The signals may then be modified and combined in such a way that a series of pulses is generated for only one direction of motion. A similar set of pulses corresponding to the reverse direction may be generated by a different combination and the two can be combined to provide a directional system of counting (Shepherd 1963).

The method by which pulses are achieved is summarized in Figure 9.3. Square waveforms A and C are derived from the sinusoidal signals by amplifying and limiting and are inverted to give signals B and D. Two series of pulses A' and B', corresponding to the steep edges of the square waveforms are then obtained by differentiating A and B. These are combined with the square waveforms and a bias applied so that only the combinations of a positive pulse and a positive square wave gives an output. Whether this occurs depends upon the sign of the phase difference between the input waveforms A and B, and this depends upon the direction of travel. By differentiating waveforms C and D and combining in a similar way, it is possible to generate four pulses for each cycle so that one digit is counted for a movement of one quarter of the grating spacing. In this way the fringes are in effect subdivided by

[†] Where θ is the angle between the gratings.

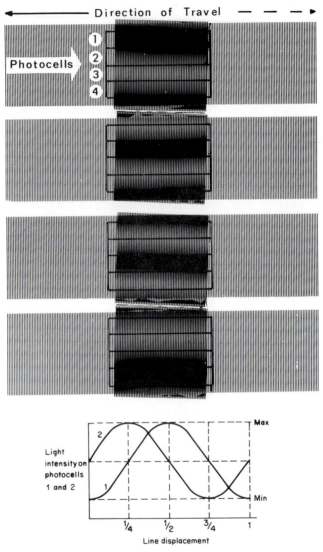

Figure 9.2 Moiré fringes and light distribution. (After Shepherd 1963.)

a factor of 4. Other systems have been devised which carry subdivision still further, in some cases up to a factor of a thousand, using either digital or analogue methods so that very precise measurements may be made with compartively coarse gratings (Macilraith 1964).

In order to avoid damaging the gratings as they move past each other it is necessary that they are separated by a small gap and this

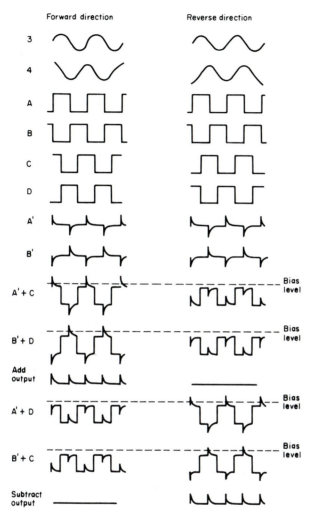

Figure 9.3   Derivation of pulse output from quadrature waveforms. (After Shepherd 1963.)

is liable to reduce the contrast of the moiré fringes. For coarse gratings one need only consider the geometrical shadow of one grating cast on the other. The effect of the gap depends upon the grating spacing (pitch) and the angular subtense of the illumination. From Figure 9.4 it is evident that if the angular subtense of the source is ϕ and the period of the grating is d then at a distance $g = d/4\phi$ for an equal-bar-space grating and at a distance $d/2\phi$ for a

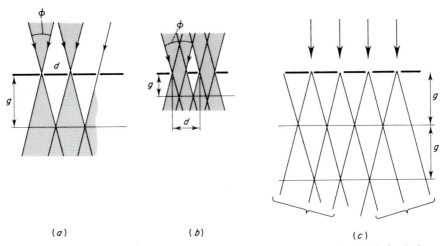

(a)   (b)   (c)

Figure 9.4 The effect of the gap between the measuring grating and the index piece.

grating with thin lines, the contrast will be zero. For gratings of finer pitch it is necessary to take into account the effects of diffraction. When a grating is illuminated by a plane monochromatic wavefront, as depicted in Figure 9.4(c), it forms so-called "Fresnel images" of itself in a series of planes parallel to the grating and separated by a distance $g = nd^2/2$, where n is an integer. When n is even, the images are in phase with the grating; when n is odd, they are out of phase by π, so that a dark region on the grating is opposite a light region on the image. In order to achieve maximum contrast of the moiré fringes it is necessary to ensure that if the index grating does not lie in contact with the scale grating then it must lie in the plane of one of the Talbot images. For monochromatic light a large number of images may be observed, but for broad-band illumination the higher-order images soon become indistinct and in practice it is usually necessary to work in the plane of the first-order image.

For very fine gratings this creates mechanical problems because the gap is inversely proportional to the square of the pitch. For a 250 line mm^{-1} grating used with radiation of wavelength 0.85 μm (the peak of sensitivity of a typical silicon cell), the gap would be 19 μm. For a 1000 line mm^{-1} grating it would be only 1.2 μm. To maintain the latter constant to, say, 20% would require the gratings to be made on surfaces which were flat to within ± 0.24 μm and would require extremely clean conditions so that no dust or grit could find its way between the gratings. Because of this problem, most systems requiring high precision have used comparatively coarse gratings with analogue subdivision of the signal rather than gratings of fine pitch. However, a digital output is often very much more convenient, particularly if the system is to be used in conjunction with numerically controlled machinery or if the output is to be fed into a digital computer. From this point of view, finer gratings are to be preferred in a system which gives one pulse per increment of measurement. If the resolution is to be, let us say, $1\,\mu m$ or less, then this requires gratings with a period of $4\,\mu m$ or less and a gap of $20\,\mu m$ or less. This is difficult to achieve but not impossible, at least over small distances, and a grating of this type has been used as the basis for a hand-held digital micrometer.

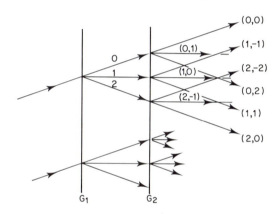

Figure 9.5 Diffracted orders from a pair of gratings.

One system which was designed for use with fine gratings, particularly blazed transmission gratings, is the so-called "spectroscopic reading head" (Burch 1963). Nowadays it is little used because most metrological gratings are of the bar-and-space variety, but it is of interest because it illustrates a slightly different approach to the formation of moiré fringe patterns. In Figure 9.5 a collimated beam of light is split by the first grating G_1 into a series of diffracted orders which we label 0, 1, 2, etc. Each order is then incident upon the second grating and is in turn split into a series of diffracted orders so that there emerges a series of wavefronts in certain discrete directions. If the gratings are parallel and of equal spacing then a wavefront may emerge at a given angle by a variety of routes. If we label the wavefronts according to their orders of diffraction at the two gratings then for example wavefronts $(0, 2)$, $(1, 1)$, $(2, 0)$,

(3, − 1) all emerge in the same direction. They will therefore interfere to give a uniformly illuminated field of which the intensity depends upon the relative phases between the contributions diffracted in the different ways. This depends upon the relative phase of the two gratings, and will vary periodically as one is moved relative to the other. If there is any misalignment between the two gratings then the directions of the various wavefronts will not coincide and the field of view will be crossed by wavefront interference fringes which are the same as the moiré fringes observed between the two gratings. The two sets of fringes are equivalent and indeed it is usually possible to explain moiré fringes in terms of wavefront interference. In this case, for example, it is equally valid to argue that the two gratings generate a moiré pattern and this is then superimposed on any wavefront diffracted from the grating, or alternatively that the various wavefronts emerging from this second grating interfere to produce the fringe pattern. In the case of the spectroscopic reading head the shape of the grooves is chosen in such a way that most of the energy goes into only two of the diffracted wavefronts which emerge at the same angle as the incident radiation (e.g. orders (0, 1) and (1, 0). This ensures that the fringes have the form of those from two-beam rather than multiple-beam interferometry and are therefore simpler, and more easily analysed.

It is possible to design a system which is insensitive to the gap between the measuring grating and the index, but in effect this uses three gratings rather than just two. Just as a grating will generate images of itself at various distances from its surface, so it may also be used to form an image of another grating, provided that certain conditions relating the pitches of the gratings to the image and object distances are met. As before one can differentiate between those conditions when geometrical optics may be applied and those when the imaging properties are determined by diffraction. The conditions for image formation are summarized in Figure 9.6 (Pettigrew 1977). In this type of system moiré fringes are obtained by placing a third grating of the appropriate pitch in the position of the image of the first grating. As the image-forming grating is moved, the image moves by a greater distance, so that the sensitivity is magnified by an "optical lever" effect. In practice it would be unnecessarily cumbersome to use three separate gratings and the tolerance on the depth of focus would be no less severe than for a two-grating system. However, if a reflection grating is used to form a diffracted image, then it is possible to arrange that the moiré pattern is formed between the object grating and its own image and the two remain exactly

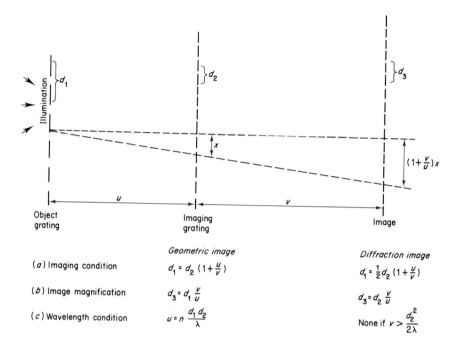

Figure 9.6 in the diagram:

Illumination

d_1 d_2 d_3

x

$(1+\frac{v}{u})x$

Object grating   Imaging grating   Image

u v

	Geometric image	Diffraction image
(a) Imaging condition	$d_1 = d_2 \left(1+\frac{u}{v}\right)$	$d_1 = \frac{1}{2}d_2\left(1+\frac{u}{v}\right)$
(b) Image magnification	$d_3 = d_1\frac{v}{u}$	$d_3 = d_2\frac{v}{u}$
(c) Wavelength condition	$u = n\frac{d_1 d_2}{\lambda}$	None if $v > \frac{d_2^2}{2\lambda}$

Figure 9.6 Image formation in a three-grating system.

superimposed over a very wide range of gaps. The measuring system then consists of a reflecting scale grating of groove spacing d and a measuring head or transducer consisting of a source, a detector and a grating of pitch d, for diffraction imaging, or $2d$ if the geometrical image is used. The fringe contrast depends upon the distribution of light among the various diffracted orders from the imaging grating and therefore is a function of the groove profile of that grating. The contrast for the geometrical image is better than for a diffracted image but is more sensitive to the size of the gap. It varies from nearly 100% to zero as the gap is increased, whereas for the diffracted image it is constant, provided that $g > d^2/2\lambda$. As we saw in Chapter 4, this is the system that Gerasimov used for controlling the position of the grooves on his ruling engines for the manufacture of spectroscopic gratings.

☐ The manufacture of metrological gratings

The earliest gratings that were used for metrological purposes were printers' screens. In most cases these were not of sufficient accuracy and it was only when resin replicas from the Merton–NPL process became available that moiré fringe measuring systems developed to the stage where they could be applied, for example, to machine tool control. These gratings were of a form intended for use as spectroscopic reflection gratings in the infrared. They therefore had sawtooth groove profiles, but were not coated with a reflecting layer since they were to be used in transmission. The best contrast is obtained with gratings consisting of alternate opaque and transparent strips of equal width and it is in this form that most metrological gratings are made. The most common is the two-grating system employing gratings of which the pitch is between 10 and 250 lines mm^{-1}. These are rather coarser than most spectroscopic gratings, except for those used in the infrared, although with the "three-grating" system finer gratings may be used.

The width of the lines is therefore usually $2\,\mu$m or greater and lies within the range of conventional photographic and lithographic techniques. In most cases the gratings are produced originally on photographic plates and then copied onto photoresist and ultimately onto chromium on glass, using the same techniques that we have described in Figure 8.12 for the production of X-ray gratings. The original grating may be copied photographically from a ruled grating, it may be a reduction of a larger drawing, or it may be a photograph of interference fringes (indeed, it was for this purpose that the first interference gratings were made by Burch in the early 1960s). Frequently only a small area of grating is made in the first instance and this is then transferred many times in register onto a longer plate. During the process the quality of the grating may be improved because a given area receives several exposures to different parts of the master. The position of the lines is then an average of the various exposures and errors in positions of the lines on the master are averaged out. This is effectively the optical equivalent of the pith nut in the Merton–NPL process. The copying system is shown in more detail in Figure 9.7 (Purfit *et al.* 1974).

A is a long grating made by a previous process (e.g. by joining several shorter gratings), B is a photographic plate of the same size and both A and B are mounted end-to-end on a slowly moving carriage. C is a fixed reading head which passes collimated light through both an index grating and grating A and produces an alternating signal from a photodetector. These signals are then used to trigger a xenon flash lamp on the fixed printed head D, thus exposing the photographic plate to the pattern defined by the printing index. In this way grating A is transferred into B with two processes of

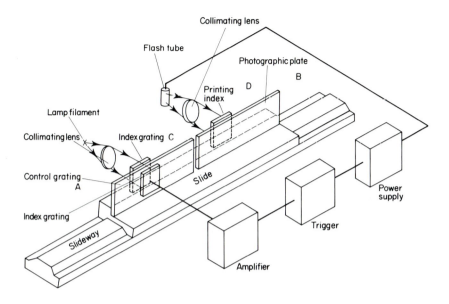

Figure 9.7 Equipment for the copying and correction of metrological gratings. (After Purfit *et al.* 1974.)

averaging. First the photocell output averages the moiré fringe pattern from a large number of lines of grating A, and second each line on B is the cumulative result of exposure to every line on the printing index. Errors in both grating A and the printing index may therefore be averaged out to the extent that the residual errors are determined by the stability of the material on which the grating is made and the effects of temperature variations during the copying process. In this way it is possible to make gratings up to 1 m long in which the departure from linearity nowhere exceeds 0.5 μm. For some applications even longer gratings are required and these may be produced by joining together a series of shorter gratings. In so doing it is important that the different gratings should remain in phase; this is achieved by adjusting the relative position of two components with respect to the moiré pattern between them and a third grating which bridges the join. It does not matter if a few lines are missed at the join because in use the moiré signal will be averaged over a comparatively large area, but if the relative phase is not properly adjusted errors will occur in the measurement.

All that has been said so far can be applied equally well to the measurement of angular displacement. For angular measurement a radial grating is used in which the lines are arranged like the spokes of a wheel. Usually the grating takes the form of an annulus between 10 and 20 mm wide on a wider annular blank. A grating with an accuracy of ± 1 second of arc can be made with a conventional circular dividing engine by the multiple printing of a small segment of grating using the same flashing technique as used for making linear gratings. Residual errors may then be reduced still further by multiple printing the whole of the grating at all orientations, again using the flashing technique, and in this way gratings have been made to an accuracy of 0.1 seconds of arc. Radial gratings are available with up to 43 200 lines which corresponds to one line per 0.5 minute of arc and more precise measurements are achieved by fringe interpolation. Such gratings are usually used with two reading heads which are diametrically opposite each other. In this way it is possible to eliminate the effects of centring errors.

□   Measurement of strain and measurement of form

In addition to the measurement of displacement of one grating with respect to another it is also possible to use gratings in the measurement of the shape and the change of shape of both regular or irregular objects. One of the simplest methods of achieving this is to illuminate an object with collimated light through a coarse bar and space grating so that the shadows of the lines delineate the contours of the object. Figure 9.8(a) illustrates the principle of this technique and an example of its application. (Takasaki 1970, 1973, Welsh 1980). In this case the separation of the projected contour lines is the same as the spacing of the grating lines. A variation of the technique is shown in Figure 9.8(b) and involves a rather finer grating. In this case the object and the projected contour pattern are viewed through the grating and the shape is displayed by the moiré pattern between the distorted projection of the grating and the grating itself. The sensitivity of this technique varies with the angle between the direction of illumination and the direction of view and tends to zero if one views along the direction of illumination.

Grating techniques may also be applied to the measurement of strain or displacement in the plane of the surface of the object. Various methods are available and two of these are shown schematically in Figure 9.9.

In the method shown in Figure 9.9(a), an interference grating is recorded in a layer of photoresist which has been sprayed onto the

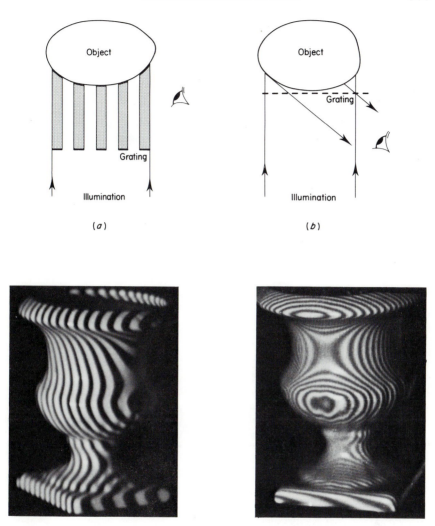

Figure 9.8 The measurement of form by (a) fringe projection, (b) moiré topography.

object (Cook 1971). The object may then either be distorted and a second grating recorded, or the first grating may be developed and the object restored to its original position. In the first case the two recorded gratings form a moiré pattern which gives a permanent record of the strain. In the second case a moiré pattern is formed

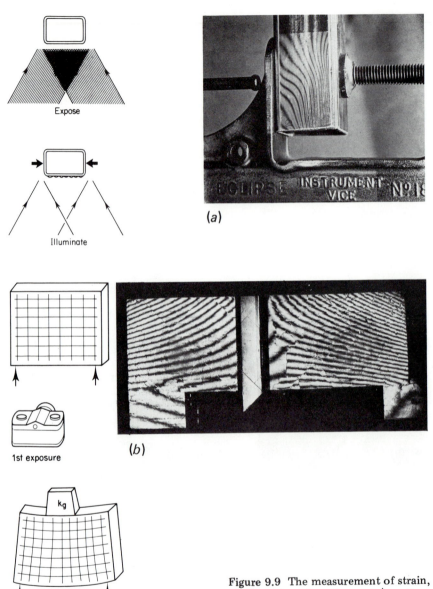

Expose

Illuminate

(a)

1st exposure

(b)

kg

2nd exposure

Figure 9.9 The measurement of strain, (a) by observing the moiré pattern between an interference grating on the surface and the fringe pattern that generated it (Cook 1971), (b) by double-exposure photography of a grating using a specially adapted camera. (Burch and Forno 1975.)

between the recorded grating and the interference fringe pattern used to record it and the strain may be viewed "live".

In the technique shown in Figure 9.9(b), a periodic pattern is applied to the object, for example, on a small scale by photolithography or on a large scale by spraying through a periodic mask. This is then photographed with a camera which has been adapted with a specially designed slotted aperture. The effect of this mask is to tune the frequency response of the camera so that for spatial frequencies near 300 lines mm^{-1} it gives almost perfect performance — far better than if the whole of the camera lens were used to form the image. A double exposure photograph is then taken, one exposure with the object unstrained and the other with it distorted and the strain is rendered visible by the moiré between the two patterns (Burch and Forno 1975).

□   The use of a grating as a beam splitter

In most of the applications that we have considered so far only one diffracted order of the grating has been used; in spectroscopy it is of course the change of angle of diffraction with wavelength that is of greatest importance. There are occasions when it is convenient to use a grating as a beam splitter to divide the incident wavefront into two or more components. The main advantage of using a grating rather than, for example, a semi-reflecting mirror is that it may be used completely in reflection and thus can be used at wavelengths for which transmitting materials are not available or at power levels where the absorption by a conventional beam splitter would lead to damage. An example of the latter is in the study of very high power laser beams where it is required to "siphon off" a small proportion of the beam for diagnostic purposes. In this case what is required is a very inefficient grating where the efficiency in the first order is perhaps only a fraction of one percent and does not detract significantly from the power in the main beam (Loewen et al. 1976). In other applications the beams need to be of more equal intensity.

Figure 9.10 shows, for example, a grazing incidence interferometer for the study of the flatness of non-optical surfaces. Here, two gratings are used, one to split the beam and the other to recombine them. This particular design is due to Birch (1973), but others have employed similar principles. The first grating divides the wavefront into two components, the zero order is undisturbed while the first

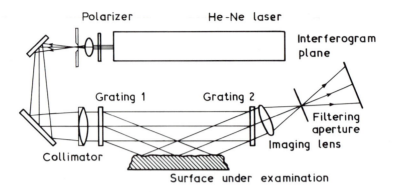

Figure 9.10 Schematic diagram of an oblique incidence interferometer employing a diffraction grating as a beam splitter.

diffracted order is reflected obliquely by the surface under examination. This is then recombined with the zero order at the second grating to generate interference fringes which are characteristic of the shape of the surface under test. At grazing incidence the optical path difference introduced by a departure from flatness of Δh is (as we saw in Chapter 8) equal to $\Delta h \cos i$ and the change in order of interference is

$$\Delta N = \frac{2\Delta h \cos i}{\lambda}$$

However, the angle of incidence is determined by the grating equation and in this case

$$\cos i = \sin \theta = \lambda/d$$

so that

$$\Delta N = 2\Delta h/d$$

At normal incidence as, for example, in a Fizeau interferometer, the corresponding equation is

$$\Delta N = 2\Delta h/\lambda$$

so that at normal incidence one fringe corresponds to a departure of flatness of half a wavelength. At grazing incidence one fringe corresponds to one half the grating period. The fringe pattern may then be interpreted in the same way as a Fizeau interferogram with an equivalent wavelength equal to the grating period. An interferometer of this type is particularly useful for the study of engineering surfaces, first because the sensitivity may be chosen more or less at will (whereas normal incidence interferometers are generally too sensitive); second because at grazing incidence adequate reflectance is achieved from non-optical surfaces and it is therefore possible to study even ground or machined surfaces; and third, because of the obliquity, it is possible to study areas which are between about 5 and 20 times wider than the gratings. If the gratings are approximately square the area of the sample that is illuminated is a long strip. However, by correlating the measurements from a series of strips it is possible to study larger areas and, for example, an interferometer employing gratings 70 mm wide (groove length) by 100 mm high has been used for such tasks as testing the flatness of granite surface tables.

☐   Wire grid polarizers

A grid of parallel conducting wires will act as a polarizer for electromagnetic radiation of which the wavelength is long compared with the spacing of the grid. For radiation with its electric vector parallel to the wires the grid behaves as a conductor and reflects, but if the electric vector is perpendicular to the wires then the grid behaves as an insulator and the radiation is transmitted, as shown in Figure 9.11(a). For long-wavelength radiation, such as microwaves and radio waves, the manufacture of suitable grids is straightforward because the scale of the device is so large that unsupported wires may be used. The finest unsupported wire grids that have been made of any appreciable size have wires of a diameter of 5μm and are suitable for wavelengths longer than 40μm (Costley *et al.* 1977). For shorter wavelengths the wires have to be supported on a transparent substrate such as Mylar or polythene and are produced *in situ* by methods of conventional photolithography (Auton 1967). A contact printing mask is produced by the step and repeat photographic reduction of a drawing of a grid. The final wires are made by etching through the photoresist mask into a metal coating on the substrate. The finest grids that may be made this way have a spacing of 4μm and are effective as polarizers for wavelengths greater than about 20μm.

For wavelengths below 1μm, sheet polaroid or polarizing crystals may be used. For the region between 1μm and 20μm it used to be necessary to employ alternative methods such as Brewster angle reflection devices or piles of plates. These are both bulky and sensitive to angle of incidence and therefore difficult to use in commercial spectrophotometers where space is limited and where divergent

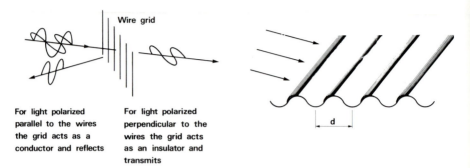

For light polarized parallel to the wires the grid acts as a conductor and reflects

For light polarized perpendicular to the wires the grid acts as an insulator and transmits

d

Figure 9.11 Wire grid polarizers.

light is often used. This problem was overcome very simply by using a diffraction grating formed on a material which is transparent in the infrared (for example germanium, calcium fluoride or KRS5). This may then be coated with aluminium by vacuum deposition in such a way that only the tops of the grooves are coated, as shown in Figure 9.11(*b*). Originally, plastic replicas or gratings ruled directly into the substrate were used, but plastics have strong absorption bands and the master rulings are very expensive (Bird and Parrish 1960). The use of interference gratings has overcome these problems. The thin layer of photoresist absorbs very little (or may be removed completely using the methods shown in Figures 8.11 and 8.12) and grids with periods as small as $0.22\,\mu m$ have been made in this way. Their performance in the infrared is excellent, as shown in Figure 9.12, and they show strong polarizing effects even in the visible.

☐   Grating couplers for optical waveguides

Recently there has been a great deal of interest in the development of optical waveguides which consist of thin layers of dielectric material coated on a suitable substrate. The light wave is constrained to travel in these layers by total internal reflection as in an optical fibre, and it is possible to build in this way integrated optical circuits which have the advantage over conventional electronics of greater compactness and speed. However, one problem which arises is that of coupling the light in to and out of these devices. A robust and convenient solution to the problem was found by preparing on the surface of the waveguide a small interference diffraction gating

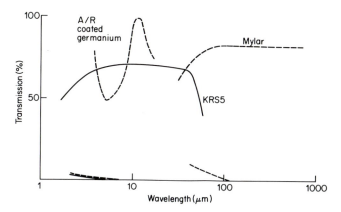

Figure 9.12 Transmission characteristics of wire grid polarizers.

(Daks *et al.* 1970). This was arranged so that the angle of diffraction exceeded the critical angle and the diffracted beam was trapped in the waveguide. A similar grating was then used to couple the light out of the device.

The conditions for coupling are similar to those for coupling light into a surface plasmon (Equation (6.7)), except that in this case the momentum of the guided wave is

$$\frac{\hbar}{2t} \cos \theta$$

where t is the thickness of the layer and θ is given by

$$m\pi = 2n\pi \frac{t}{\lambda} \cos \theta - (\theta_{12} + \theta_{23})$$

where m is the mode number, n is the refractive index, θ_{12} and θ_{23} are the phase changes on reflection at the surfaces of the layer. In practice the grating may well perturb the propagation of the mode so the calculation of the coupling conditions requires a detailed knowledge of the groove profile.

Such gratings could be made on a ruling engine, but interference methods offer the advantages of speed and flexibility. Furthermore, it is possible to prepare couplers on virtually any surface and in any shape since, if positive resist is used, the grating may be given a second exposure through a conventional microcircuit mask. The production of integrated circuits will almost certainly entail conventional photofabrication and it therefore seems logical to apply interference technique to the production of couplers.

Appendix

The Choice of a Grating

The performance of a grating is influenced by a great many factors. First there are the theoretical factors we discussed in Chapter 2, which determine the performance of an ideal grating. Secondly there are practical factors, in which the technology of grating manufacture imposes its own limits upon the performance that can be achieved — for example, the total length of groove that can be ruled, the loss of resolution due to the departure from flatness of the wavefront or spectral noise introduced by grating imperfections. Thirdly there are factors that depend upon the way in which a grating is to be used — for example, use over a wide range of wavelengths will involve some compromise on efficiency and the relative significance of different forms of stray light will depend upon whether the grating is used in a spectrograph or a monochromator. The choice of grating for a particular application is therefore often a difficult one and in describing the relative merits of different types of grating it is almost impossible to state categorically that one is better than another. Not only does the choice depend upon the application, but it also depends on whether one is choosing the best grating for a given instrument or is choosing a grating for a given task and designing an instrument that will use to the best possible advantage the characteristics of the grating. In the former case one might well be restricted in the choice of pitch in order not to change the wavelength scale. In a monochromator designed to use a 1200 groove mm^{-1} ruled grating in the visible, the introduction of a 1200 groove mm^{-1} sinusoidal interference grating would be impractical because Wood's anomalies would occur in the middle of the spectrum. On the other

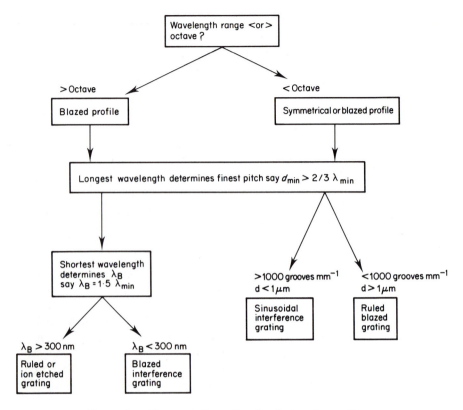

Figure A.1 Factors influencing the choice of a grating.

hand, an $1800\,\mathrm{groove\,mm}^{-1}$ interference grating would avoid this, would have a similar efficiency and a very much lower level of stray light and the extra dispersion would, as we see from Equation (3.6), increase the transmission of the instrument by 50%. In a spectrograph, though, unless the instrument were designed specifically to take advantage of the increased dispersion, the introduction of a finer grating may well be a net disadvantage.

In view of the wide range of forms of spectroscopy, it is not possible simply to list the various types of grating that are available and the applications to which they are best suited. The final choice of a grating must be made individually after all relevant factors have been taken into account. In Figure A.1 we summarize some of the points which have to be considered in making this choice.

First of all one must consider the range of wavelengths for which the grating is intended. In general if it is greater than an octave, then a grating with a blazed saw-tooth profile is required. There is a better chance of avoiding serious anomalies than with a symmetrical profile and the overall efficiency tends to be rather higher. Next one needs to consider the longest wavelength, because this will determine the finest pitch and hence the greatest free spectral range that can be obtained. As a rule one would not expect the groove spacing to be less than about $0.6\lambda_{max}$, although availability and cost may well suggest that it be greater than this.[†] Having chosen the pitch, the choice of blaze wavelength is determined mainly by the shortest wavelength in the range. We saw in Chapter 2 that ideally the efficiency falls to zero at $\frac{1}{2}\lambda_b$ because at that wavelength all of the energy should be in the second order. The efficiency then falls to about half its maximum value at about $\frac{2}{3}\lambda_b$. If we accept this as the condition at the shortest wavelength in the range, then it would follow ideally that $\lambda_{min} > \frac{2}{3}\lambda_b$. The choice between a ruled and an interference grating will largely depend upon the value of λ_b. If λ_b is less than 250 nm, then an interference grating is probably the best choice because it has the same efficiency characteristics as the ruled grating but a much lower level of stray light. At longer blaze wavelengths, blazed interference gratings are not always available, so a ruled grating may be the best choice. However, this may depend upon the relative importance of efficiency and spectral purity; for some applications it may be better to use a less efficient interference grating than a more efficient ruled grating with a high level of stray light.

If the wavelength range is less than an octave, then one may use either a saw-tooth or a symmetrical profile and the choice will probably depend upon the dispersion that is required. We have seen that a sinusoidal groove profile is as efficient as the equivalent blazed profile for wavelengths between $0.6d$ and $1.65d$; that is, when the blazed facet angle is greater than about $20°$. This corresponds to conditions of fairly high dispersion and under these circumstances one would probably choose an interference grating rather than a ruled one, in order to profit from the improvement in stray light and the fact that interference gratings of fine pitch are more widely available than ruled ones. In this case one would choose the pitch to be about $0.6\lambda_{max}$. On the other hand, if there is some constraint on the

[†] It is at this stage that one must decide whether to use an echelle or a first-order grating. As we have seen, the echelles are particularly appropriate for spectrographs in which a wide range of wavelengths is recorded, usually at high resolution. First-order gratings are used in monochromators where one wishes to avoid the overlapping of orders.

dispersion as might be imposed, for example, by the field angle of the camera optics in a spectrograph, then a coarser grating may be required. In this case one must choose between the greater efficiency and control over anomalies that is afforded by a ruled grating and the lower efficiencies but purer spectrum of an interference grating. If the blaze wavelength is less than about 250 nm, then a blazed interference grating offers the best of both worlds. If the blaze wavelength is greater than 250 nm and the pitch is coarser than about 1000 grooves mm^{-1}, then a ruled grating is usually to be preferred. Thus we see that the more recent interference techniques for making gratings and the classical ruling technique are to a large extent complementary. In many cases interference gratings have replaced ruled gratings and in some cases it would be impossible to produce a suitable ruled grating. Nevertheless there are areas in which interference techniques are unsuitable and one has to use a ruled grating. This is particularly so for echelles and for coarse gratings such as those used in the infrared, but even for the visible and near ultraviolet there remain applications for which the ruled grating is the best choice.

Epilogue

☐   Decorative applications

Throughout the course of this book we have studiously considered the operation and use of diffraction gratings. We have described their workings in terms of physical optics and in some cases in terms of quantum mechanics. They have revolutionized spectroscopy and as such have contributed by an immeasurable amount to man's knowledge and understanding of the world in which he lives. It is this which justifies the continued effort in developing the science and the technology of the diffraction grating.

The fact remains that a grating splits light up into its constituent *colours*. Anyone who has seen or handled a good optical grating has been impressed by the intensity and purity of the colours he sees, particularly when the grating is viewed under optimum conditions. Although none would deny the scientific importance of a grating, the sheer beauty of the device cannot be ignored and throughout the history of the development of gratings this factor must surely have offered some consolation to the frustrated worker seeking to achieve perfection for scientific reasons.

It is perhaps surprising that the use of gratings as objects of decoration and ornament has not been taken up more widely. Although occasional items of jewellery have been produced, no doubt in idle moments using reject pieces of replica, it is only very recently that anything of this nature has become widely available. It is now possible to obtain arrays of gratings embossed into aluminized plastic sheet. These are usually circular gratings having spiral grooves like that of a gramophone record but with approximately $600\,$grooves$\,mm^{-1}$. The advantage of this is that the observation of a spectrum depends less critically upon the direction of illumination, since there will always be some part of the grating where the grooves have the correct orientation. Furthermore as one's direction of view changes, so does the area from which the spectrum appears to come and this gives rise to a scintillating effect. Applications of this type of diffracting foil range from the gaudy but eye catching use in advertising displays to the (occasionally) more tasteful incorporation into costume jewellery.

A rather more durable ornament of rather better optical quality may be achieved by forming an interference grating in the normal way, but by exposing the photoresist to an incoherent light source through a mask before development. In this way the mask protects certain regions of the grating but allows others to be completely overexposed. On development the overexposed resist is removed and a grating is left only in those areas protected by the mask. When the grating is aluminized it has the appearance of a plane mirror but with the pattern of the mask visible in the colours of the spectrum. Because of the high spatial resolution of photoresist, very fine detail may be incorporated and the technique is particularly effective with coats of arms or other insignia.

Whether or not the decorative aspects of diffraction gratings should be taken seriously is a matter for speculation. There is no doubt that the main importance of gratings stems from their ability to disperse radiation into a spectrum that can be analysed, and that has been the subject of this book. However, it may be fitting to finish more or less where we began, with Barton's Buttons, and to note that the study of diffraction gratings offers more reward than the mere achievement of scientific results. Those involved in the production of gratings may well derive encouragement from the words of Horace:[†] "Omne tulit punctum qui miscuit utile dulci" — "He gains everyones approval who mixes the pleasant with the useful".

[†] *Ars Poetica*, 343; quoted in "Harvest of a Quiet Eye" by A. L. Mackay, 1977, The Institute of Physics, Bristol and London.

References

Anderson, W. A., Griffin, G. L., Mooney, C. F. and Wiley, R. S. (1965). *Appl. Opt.* **4**, 999.

American Society for Testing and Materials (1969). *Yearbook* E387-69T.

Auton, J. P. (1967). *Appl.* **6**, 1023.

Auton, J. P. and Hutley, M. C. (1972). *Infrared Phys.* **12**, 95.

Ayoagi, Y. and Namba, S. (1976). *Optica Acta* **23**, 701.

Babcock, H. W. (1962). *Appl. Opt.* **1**, 417.

Babcock, H. D. and Babcock, H. W. (1951). *J. Opt. Soc. Am.* **41**, 776.

Bartlet, I. R and Wildy, P. C. (1975). *Appl. Opt.* **14**, 1.

Barton, J. N. (1822). UK Patent No. 4678.

Bennett, H. E. and Porteus, J. O. (1961). *J. Opt. Soc. Am.* **51**, 123.

Bennet, J. M. (1900). *J. Phys. E.* **2**, 816.

Beutler, H. G. (1945). *J. Opt. Soc. Am.* **35**, 311.

Birch, K. G. (1966). *J. Sci. Instrum.* **43**, 243.

Birch, K. G. (1973). *J. Phys. E.* **6**, 1045.

Bird, G. R. and Parrish, M. (1960). *J. Opt. Soc. Am.* **50**, 886.

Boivin, L. P. (1973). *Opt. Commun.* **9**, 206.

Bottema, M. (1980). *Proc. Conf. on Gratings and Period Structures, San Diego. S.P.I.E.* **240**, 171.

Brandes, R. G. and Curran, R. K. (1900). *Appl. Opt.* **10**, 2101.

Breidne, M., Johansson, S., Nilsen, L. E. and Ahlen, H. (1979). *Optica Acta* **26**, 1427.

Breidne, M. and Maystre, D. (1980). *Appl. Opt.* **19**, 1812.

Bruner, E. C. (1972). *Appl. Opt.* **11**, 1357.

Bryngdahl, O. (1970). *J. Opt. Soc. Am.* **60**, 140.

Burch, J. M. (1960). *Research* **13**, 2.

Burch, J. M. (1963). "Progress in Optics" (ed. E. Wolf), Vol. II, p. 75. North Holland, Amsterdam.

Burch, J. M. and Palmer, D. A. (1961). *Optica Acta* **8**, 73.

Burch, J. M. and Forno, C. (1975). *Opt. Engng.* **14**, 178.

Burton, W. M. and Reay, N. K. (1970). *Appl. Opt.* **9**, 1227.

Butler, D. (1973). *Micron* **4**, 410.

Cerutti-Maori, G. and Petit, R. (1970). *Nouv. Rev. d'Opt. Appl.* **1**, 321.

Clapham, P. B. and Dew, G. D. (1967). *J. Sci. Instrum.* **44**, 899.

Compton, A. H. and Doan, R. L. (1925). *Proc. Natl. Acad. Sci, USA* **11**, 598.

Cook, R. W. E. (1971). *Opt. Laser Tech.* **3**, 71.

Cordelle, J., Flammand, J., Pieuchard, G. and Labeyrie, A. (1969). "Optical Instrumentation and Techniques" (Ed. J. Home-Dickson). Oriel Press, London.

Cornu, A. (1893). *C. R. Acad. Sci. Paris* **116**, 1215.

Costley, A. E., Hursey, K. H., Neill, G. H. and Ward, J. M. (1977). *J. Opt. Soc. Am.* **67**, 979.

Cowan, J. J. and Arakawa, E. T. (1970). *Z. Phys.* **235**, 97.

Cowan, J. J., Arakawa, E. T. and Painter, L. R. (1969). *Appl. Opt.* **8**, 1734.

Daks, M. L., Kung, L., Heidrich, P. F. and Scott, B. A. (1970). *Appl. Phys. Lett.* **16**, 523.

Danielsson, A. and Lindblom, P. (1974). *Optik* **41**, 441.

Danielsson, A. and Lindblom, P. (1975). *Optik* **41**, 465.

Deleuil, R. (1969). *Optica Acta* **16**, 23.

Dravins, D. (1978). *4th Trieste Astrophys. Coll. "High Resolution Spectrometry".*

Eagle, A. (1910). *Astrophys. J.* **31**, 120.

Fischer, B., Marschall, N. and Queisser, H. J. (1973). *Surface Sci.* **34**, 50.

Franks, A., Lindsey, K., Bennett, J. M., Speer, J., Turner, D. and Hunt, D. J. (1975). *Phil. Trans. R. Soc. London* **277**, 503.

Franks, A., Lindsey, K., Bennett, J. M., Speer, J., Turner, D. and Hunt, D. J. (1977). *Sci. Prog.* **64**, 371.

Fraunhofer, J. N. (1821). *Denkschr. Kg. Akad. Wiss. Munchen* **8**, 1.

Fraunhofer, J. N. (1823). *Ann. Physik* **74**, 337.

Gale, B. (1966). *Optica Acta* **13**, 41.

Gale, H. G. (1937). *Astrophys. J.* **85**, 49.

Gee, A. E. (1975). *Jap. J. Appl. Phys.* **14** (Suppl. 1), 169.

Gerasimov, F. M. (1967). *Appl. Opt.* **6**, 1861.

Guild, J. (1956). "The Interference Systems of Crossed Diffraction Gratings". Clarendon, Oxford.

Guild, J. (1960). "Diffraction Gratings as Measuring Scales". Oxford U.P., London.

Haber, H. (1950). *J. Opt. Soc. Am.* **40**, 153.

Hägglund, J. and Sellberg, F. (1966). *J. Opt. Soc. Am.* **56**, 1031.

Hammer, D. C., Arakawa, E. T. and Birkhof, R. D. (1964). *Appl. Opt.* **3**, 79.

Hanson, W. F. and Arakawa, E. T. (1966). *J. Opt. Soc. Am.* **56**, 124.

Harada, T. and Kita, T. (1980). *Appl. Opt.* **19**, 3987.

Harada, T., Moriyama, S. and Kita, T. (1975). *Jap. J. Appl. Phys.* **14** (Suppl. 1), 175.

Harrison, G. R. (1949). *J. Opt. Soc. Am.* **39**, 522.

Harrison, G. R. and Archer, J. E. (1951). *J. Opt. Soc. Am.* **41**, 495.

Harrison, G. R. and Stroke, G. W. (1955). *J. Opt. Soc. Am.* **45**, 112.

Hessel, A. and Oliner, A. A. (1965). *Appl. Opt.* **4**, 1275.

Horsfield, W. R. (1965). *Appl. Opt.* **4**, 189.

Hutley, M. C. (1973). *Optica Acta* **20**, 607.

Hutley, M. C. (1975). *Optica Acta* **22**, 1.

Hutley, M. C. and Bird, V. M. (1973). *Optica Acta* **20**, 771.

Hutley, M. C. and Hunter, W. R. (1981). *Appl. Opt.* **20**, 247.

Hutley, M. C. and Maystre, D. (1976). *Opt. Commun.* **19**, 431.

Hutley, M. C., Verrill, J. F. and McPhedran, R. C. (1974). *Opt. Commun.* **11**, 207.

Jaquinot, P. (1954). *J. Opt. Soc. Am.* **44**, 761.

Jaquinot, P. and Rozien-Dossier, B. (1964). "Progress in Optics". (ed. E. Wolf), Vol. III. North Holland, Amsterdam.

Janot, C. and Hadni, A. (1962). *J. Phys. Rad.* **23**, 152.

JOBIN-YVON (1973). "Diffraction Gratings: Ruled and Holographic". Jobin-Yvon Company Publication, Longjumeau, France.

Johnson, L. F., Kammlott, G. W. and Ingersoll, K. A. (1978). *Appl. Opt.* **17**, 1165.

Johnson, P. D. (1957). *Rev. Sci. Instrum.* **28**, 833.

Johnson, R. L. (1975). Ph.D. Thesis, London University.

Jovecivic, S. and Sesnic, S. (1976). *Optica Acta* **22**, 461.

Kalhor, H. A. and Neureuther, A. (1971). *J. Opt. Soc. Am.* **61**, 43.

Labeyrie, A. (1967). *Proc. Conf. Optics, Marseille Centre Nat. d'Etudes Spatiales*, Report No. 00015/PR/ED.

Labeyrie, A. and Flamand, J. (1969). *Opt. Commun.* **1**, 5.

Learner, R. C. (1972). *Kitt Peak Natl. Observatory*, A.U.R.A. report No. 39.

Lindsey, K. and Penfold, A. B. (1976). *Opt. Engng.* **15**, 220.

Loewen, E. G. (1972). *Opt. Spectra* **6**(4), 32.

Loewen, E. G., Nevière, M. and Maystre, D. (1976). *Appl. Opt.* **18**, 4178.

Loewen, E. G., Nevière, M. and Maystre, D. (1977). *Appl. Opt.* **16**, 2711.

Lyman, T. (1901). *Phys. Rev.* **12**, 1.

Macilraith, A. (1964). *J. Sci. Instrum.* **41**, 34.

Mack, J. E., Stehn, J. R. and Edlèn, B. (1932). *J. Opt. Soc. Am.* **22**, 245.

Mack, J. E., Stehn, J. R. and Edlèn, B. (1933). *J. Opt. Soc. Am.* **23**, 184.

Madden, R. P. and Strong, J. (1958). "Concepts in Classical Optics". Freeman, San Francisco.

Maréchal, A. (1958). *Optica Acta* **5**, 70.

Maystre, D. (1973). *Opt. Commun.* **8**, 216.

Maystre, D. (1974). Thesis, Université D'Aix, Marseille.

Maystre, D. and McPhedran, R. C. (1974). *Opt. Commun.* **12**, 00.

Maystre, D. and Nevière, M. (1977). *J. Opt. (Paris)* **8**, 165.

Maystre, D. and Petit, R. (1976). *Opt. Commun.* **17**, 196.

Maystre, D., Nevière, M. and Vincent, P. (1978). *Optica Acta* **25**, 905.

McPhedran, R. C. (1973). Ph.D. Thesis, University of Tasmania.

McPhedran, R. C. and Maystre, D. (1974). *Optica Acta* **21**, 413.

McPhedran, R. C. and Waterworth, M. D. (1972). *Optica Acta* **19**, 877.

McPhedran, R. L., Wilson, I. J. and Waterworth, M. D. (1973). *Opt. Laser Technol.* **5**, 166.

Meecham, W. C. (1956). *J. Appl. Phys.* **27**, 361.

Merton, T. (1950). *Proc. R. Soc. London, Ser. A* **201**, 187.

Michels, D. (1974). *J. Opt. Soc. Am.* **64**, 662.

Michels, D. J., Mikes, T. L. and Hunter, W. R. (1974). *Appl. Opt.* **13**, 1223.

Michelson, A. A. (1915). *Proc. Am. Phil. Soc.* **54**, 137.

Michelson, A. A. (1927). "Studies in Optics". Chicago U.P.

Millar, R. F. (1969). *Proc. Camb. Phil. Soc.* **65**, 773.

Millar, R. F. (1971). *Proc. Camb. Phil. Soc.* **69**, 217.

Namioka, T. (1959). *J. Opt. Soc. Am.* **49**, 446, 460, 951.

Namioka, T., Noda, H. and Seya, M. (1973). *Sci. Light.* **22**, 77.

Namioka, T., Noda, H. and Seya, M. (1900). *J. Spectrosc. Soc. Japan* **23** (Suppl. 1), 29.

Neureuther, A. R. and Hagouel, P. I. (1974). *Electron and Ion Beam Science and Techniques: 6th International Conference.*

Nevière, M. and Hunter, W. R. (1980). *Appl. Opt.* **19**, 2059.

Noda, H., Namioka, T. and Seya, M. (1974). *J. Opt. Soc. Am.* **64**, 1031, 1037, 1043.

Onaka, R. (1958). *Sci. Light.* **7**, 23.

Palmer, C. H. (1952). *J. Opt. Soc. Am.* **42**, 269.

Palmer, C. H. and Lebrun, H. (1972). *Appl. Opt.* **11**, 907.

Palmer, C. H. and Phelps, F. W. (1968). *J. Opt. Soc. Am.* **58**, 1184.

Pavageau, J. and Bousquet, J. (1970). *Optica Acta* **17**, 469.

Petit, R. (1963). *C. R. Acad. Sci. Paris* **257**, 2018.

Petit, R. (1963). *Rev. d'Opt.* **42**, 263.

Petit, R. (1965). *Appl. Opt.* **4**, 1551.

Petit, R. and Cadilhac, M. (1966). *C. R. Acad. Sci. Paris* **262**, 468.

Petit, R. and Maystre, D. (1972). *Rev. Phys. Appl.* **7**, 427.

Pettigrew, R. M. (1977). *Proc. 1st European Congress on Optics Applied to Metrology S.P.I.E.*, Vol. 136, p. 325.

Pockrand, I. (1976). *J. Phys. D.* **9**, 2433.

Pouey, M. (1973). *C. R. Acad. Sci. Paris* (B) **277**, 459.

Pouey, M. (1974). *J. Spectrosc. Soc. Japan* **23** (Suppl. 1), 67.

Preston, R. C. (1970). *Optica Acta* **17**, 857.

Purfit, G. L., Woodward, C. A. W. and Pettigrew, R. M. (1974). *NELEX Metrology Conf. National Engineering Lab. East Kilbride, Glasgow.*

Rasudova, G. N. and Geraimov, F. M. (1963). *Opt. Spectrosc.* 14, 215, 295.

Rayleigh (1872). *Proc. R. Soc. London* 20, 414.

Rayleigh (1874). *Phil. Mag.* XLVII, 81, 193.

Rayleigh (1907). *Proc. R. Soc. London, Ser. A* 79, 399.

Ritchie, R. H. (1957). *Phys. Rev.* 166, 874.

Ritchie, R. H. (1973). *Surface Sci.* 34, 1.

Rittenhouse, D. (1786). *Trans. Am. Phil. Soc.* 2, 201.

Roger, A. and Breidne, M. (1980). *Opt. Commun.* 35, 299.

Roger, A. and Maystre, D. (1979). *Optica Acta* 26, 447.

Rowland, H. A. (1882). *Phil. Mag.* 13, 467.

Rowland, H. A. (1893). *Phil. Mag.* 35, 397.

Rudolph, D. and Schmahl, G. (1967). *Umschau Wiss.* 67, 225.

Rudolph, D. and Schmahl, G. (1968). UK Patent No. 1261213.

Rudolph, D. and Schmahl, G. (1970a). *Optik* 30, 475.

Rudolph, D. and Schmahl, G. (1970b). *Optik* 30, 606.

Rudolph, D., Schmahl, G., Johnson, R. L. and Speer, R. J. (1973). *Appl. Opt.* 12, 1731.

Sakayanagai, Y. (1967). *Sci. Light* 16, 129.

Sawyer, R. A. (1963). "Experimental Spectroscopy". Dover, New York.

Sayce, L. A. (1953). *Endeavour* 210.

Schmahl, G. (1974). *Proc. 4th Int. Conf. Vac. UV Radiation Physics, Hamburg.* Pergamon-Vieweg, Braunschweig.

Schroeder, D. J. and Hilliard, R. L. (1980). *Appl. Opt.* 19, 2833.

Seya, M. (1952). *Sci. Light* 7, 23.

Siegbahn, M. and Magnusson, T. (1930). *Z. Phys.* 62, 435.

Siegbahn, M. and Magnusson, T. (1935). *Z. Phys.* 95, 133.

Sharpe, M. R. and Irish, D. (1978). *Optica Acta* 25, 861.

Shepherd, A. T. (1964). *In. J. Mach. Tool. Des. Res.* 3, 47.

Sheridon, N. K. (1968). *Appl. Phys. Lett.* 12, 316.

Speer, R. J. (1970). "Advances in X-ray Analysis" (ed. B. L. Henke) Vol. 13, p. 382, Plenum, New York.

Speer, R. J., Chrisp, M., Turner, D., Mrowka, S. and Tregidgo, K. (1979). *Appl. Opt.* 18, 2003. [This paper appends a translation of a paper by W. P. Linnik (1933). *C. R. Acad. Sci. URSS* 5, 210, describing the point diffraction interferometer.]

Sprague, G., Tomboulian, D. H. and Bedo, D. E. (1955). *J. Opt. Soc. Am.* 45, 756.

Stewart, J. E. and Galloway, W. S. (1962). *Appl. Opt.* 1, 421.

Stroke, G. W. (1960). *Rev. d'Opt.* 39, 291.

Stroke, G. W. (1967). "Handbuch der Physik XXIX", p. 729. Springer, Berlin.

Strong, J. (1951). *J. Opt. Soc. Am.* 41, 3.

Strong, J. (1935). *Phys. Rev.* 48, 480.

Stuart, P. R., Hutley, M. C. and Stedman, M. (1976). *Appl. Opt.* 15, 2618.

Takasaki, H. (1970). *Appl. Opt.* **9**, 1467.

Takasaki, H. (1973). *Appl. Opt.* **12**, 845.

Teng, Y. Y. and Stern, E. A. (1967). *Phys. Rev. Lett.* **19**, 511.

Torbin, I. D. and Nizhin, A. M. (1973). *Sov. J. Opt. Technol.* **40**, 192.

Tsang, W. T. and Wang, S. (1976). *Appl. Phys. Lett.* **28**, 44.

Verrill, J. F. (1973). *J. Phys. E.* **6**, 1199.

Verrill, J. F. (1975). *J. Phys. E.* **8**, 522.

Verrill, J. F. (1976). *Optica Acta* **23**, 425.

Verrill, J. F. (1978). *Optica Acta* **25**, 531.

Verrill, J. F. (1981). *Optica Acta*, in press.

Wandsworth, F. L. O. (1903). *Phil. Mag.* **6**, 119.

Welford, W. T. (1965). "Progress in Optics" (ed. E. Wolf), Vol. 4, p. 243. North Holland, Amsterdam.

Welford, W. T. (1974). "Aberrations of the Symmetrical Optical System". Academic Press, London and New York.

Welford, W. T. (1975). *Opt. Commun.* **14**, 322.

Welsh, N. (1980). *NELEX Metrology Conf. National Engineering Lab. East Kilbridge, Glasgow.*

Wheeler, C. E., Arakawa, E. T. and Ritchie, R. H. (1976). *Oak Ridge Natl. Lab. Report TM-5185.*

White, J. V. and Fraser, W. A. (1949). US Patent 246478.

Wilson, I. J., McPhedran, R. L. and Waterworth, M. D. (1973). *Opt. Commun.* **9**, 263.

Wirgin, A. (1980). *Optica Acta* **12**, 1671.

Wood, R. W. (1902). *Proc. Phys. Soc.* **18**, 396.

Wood, R. W. (1910). *Phil. Mag.* **20**, 770.

Wood, R. W. (1924). *Phil. Mag.* **48**, 497.

Yamada, Y., Hara, T. and Morita, K. (1966). *J. Jap. Soc. Mech. Engrs.* **69**, 1557.

Subject Index